SpeedPro Series

THE FORD SOHC 'PINTO' & SIERRA COSWORTH DOHC ENGINES HIGH-PERFORMANCE MANUAL

AF000652

SpeedPro Series

- 4-Cylinder Engine Short Block High-Performance Manual – New Updated & Revised Edition (Hammill)
- Aerodynamics of Your Road Car, Modifying the (Edgar and Barnard)
- Alfa Romeo DOHC High-performance Manual (Kartalamakis)
- Alfa Romeo V6 Engine High-performance Manual (Kartalamakis)
- BMC 998cc A-series Engine, How to Power Tune (Hammill)
- 1275cc A-series High-performance Manual (Hammill)
- Camshafts – How to Choose & Time Them For Maximum Power (Hammill)
- Competition Car Datalogging Manual, The (Templeman)
- Custom Air Suspension – How to install air suspension in your road car – on a budget! (Edgar)
- Cylinder Heads, How to Build, Modify & Power Tune – Updated & Revised Edition (Burgess & Gollan)
- Distributor-type Ignition Systems, How to Build & Power Tune – New 3rd Edition (Hammill)
- Fast Road Car, How to Plan and Build – Revised & Updated Colour New Edition (Stapleton)
- Ford SOHC 'Pinto' & Sierra Cosworth DOHC Engines, How to Power Tune – Updated & Enlarged Edition (Hammill)
- Ford V8, How to Power Tune Small Block Engines (Hammill)
- Harley-Davidson Evolution Engines, How to Build & Power Tune (Hammill)
- Holley Carburetors, How to Build & Power Tune – Revised & Updated Edition (Hammill)
- Honda Civic Type R High-Performance Manual, The (Cowland & Clifford)
- Jaguar XK Engines, How to Power Tune – Revised & Updated Colour Edition (Hammill)
- Land Rover Discovery, Defender & Range Rover – How to Modify Coil Sprung Models for High Performance & Off-Road Action (Hosier)
- MG Midget & Austin-Healey Sprite, How to Power Tune – Enlarged & updated 4th Edition (Stapleton)
- MGB 4-cylinder Engine, How to Power Tune (Burgess)
- MGB V8 Power, How to Give Your – Third Colour Edition (Williams)
- MGB, MGC & MGB V8, How to Improve – New 2nd Edition (Williams)
- Mini Engines, How to Power Tune On a Small Budget – Colour Edition (Hammill)
- Motorcycle-engined Racing Cars, How to Build (Pashley)
- Motorsport, Getting Started in (Collins)
- Nissan GT-R High-performance Manual, The (Gorodji)
- Nitrous Oxide High-performance Manual, The (Langfield)
- Optimising Car Performance Modifications (Edgar)
- Race & Trackday Driving Techniques (Hornsey)
- Retro or classic car for high performance, How to modify your (Stapleton)
- Rover V8 Engines, How to Power Tune (Hammill)
- Secrets of Speed – Today's techniques for 4-stroke engine blueprinting & tuning (Swager)
- Sportscar & Kitcar Suspension & Brakes, How to Build & Modify – Revised 3rd Edition (Hammill)
- SU Carburettor High-performance Manual (Hammill)
- Successful Low-Cost Rally Car, How to Build a (Young)
- Suzuki 4x4, How to Modify For Serious Off-road Action (Richardson)
- Tiger Avon Sportscar, How to Build Your Own – Updated & Revised 2nd Edition (Dudley)
- Triumph TR2, 3 & TR4, How to Improve (Williams)
- Triumph TR5, 250 & TR6, How to Improve (Williams)
- Triumph TR7 & TR8, How to Improve (Williams)
- V8 Engine, How to Build a Short Block For High Performance (Hammill)
- Volkswagen Beetle Suspension, Brakes & Chassis, How to Modify For High Performance (Hale)
- Volkswagen Bus Suspension, Brakes & Chassis for High Performance, How to Modify – Updated & Enlarged New Edition (Hale)
- Weber DCOE, & Dellorto DHLA Carburetors, How to Build & Power Tune – 3rd Edition (Hammill)

www.veloce.co.uk

First published in 1997. Reprinted 1998, 1999 & 2000. This revised and updated edition first published in 2001 (reprinted 2002, 2003, April 2006, May 2007, August 2009, February 2014 and March 2021 by Veloce Publishing Limited, Veloce House, Parkway Farm Business Park, Middle Farm Way, Poundbury, Dorchester, Dorset, DT1 3AR, England. Tel 01305 260068/fax 01305 250479/e-mail info@veloce.co.uk or www.veloce.co.uk or www.velocebooks.com.

ISBN: 978-1-903706-78-7 UPC: 6-36847-00278-7

© Des Hammill and Veloce Publishing 1997, 1998, 1999, 2000, 2001, 2002, 2003, 2006, 2007, 2009, 2014 & 2021. All rights reserved. With the exception of quoting brief passages for the purpose of review, no part of this publication may be recorded, reproduced or transmitted by any means, including photocopying, without the written permission of Veloce Publishing Ltd. Throughout this book logos, model names and designations, etc., have been used for the purposes of identification, illustration and decoration. Such names are the property of the trademark holder as this is not an official publication.
Readers with ideas for automotive books, or other transport or related hobby subjects, are invited to write to the editorial director of Veloce Publishing at the above address.
British Library Cataloguing in Publication Data – A catalogue record for this book is available from the British Library.
Typesetting, design and page make-up all by Veloce Publishing Ltd on Apple Mac. Printed and bound by CPI Group (UK) Ltd, Croydon, CR0 4YY.

THE FORD SOHC 'PINTO' & SIERRA COSWORTH DOHC ENGINES HIGH-PERFORMANCE MANUAL

DES HAMMILL

Contents

Introduction 7
Using this book **8**
Essential information **8**

Chapter 1. Problem areas 10
   Standard connecting rods 10
   Camshaft lobes and rockers 12
   Camshaft pillars 13
   Camshaft bearings 13
   Piston to valve contact 13
   Auxiliary shaft gear wear 13
   Loose sprockets 14
   Valve guides 14

**Chapter 2. Short block
components 15**
   Pistons and connecting rods 15
   Crankshaft and flywheel 20
   Clutches 21

Chapter 3. Replacement parts 23
   Pistons 23
   Piston ring sets 24
   Crankshaft bearings 24
   Camshaft bearings 24

   Camshaft spray bar 24
   Camshaft kit 24
   Gaskets 25
   Special bolts 25
   Auxiliary shaft bearing 25
   Seals 25
   Valves 25
   Valve retainers and keepers 27
   Valve stem seals 27
   Oil pump and oil pump drive 27
   Sump (oil pan) 27
   Dry sump 28
   Timing belt 28
   Camshaft and auxiliary shaft thrust
     plates 29

Chapter 4. Short block rebuild 30
   Permissible bore oversizes 30
   Repaired blocks 31
   Component inspection 31
   Block & main bearings 31
   Connecting rod checks 35
   Piston pin to connecting rod fit 36
   Checking connecting rod big end
     tunnel size 37

   Connecting rod crankshaft bearing
     tunnel resizing 38
   Aftermarket connecting rods 40
   Connecting rod bolts 40
   Connecting rod bearing crush 41
   Crankshaft 42
   Crankshaft checking 43
   Crankshaft regrinding 45
   'Check-fitting' connecting rods to
     crankshaft 45
   Building the short block 46

Chapter 5. Cylinder head valves ... 49
   Valve throat size 54
   Inlet port size (std size valves) 54
   Exhaust port size (std size valves) .. 55
   Valve throat & port modifications
     (std size valves) 56
   Large valve modifications 58

Chapter 6. Compression ratio 61
   Head planing 61
   Thinner head gasket 63
   Block planing 63
   Raised top pistons 63

'O' ringing blocks............................ 64
Compression ratio summary.......... 64

Chapter 7. Camshafts 66
Standard camshaft 66
High-performance camshafts......... 66
Choosing a camshaft...................... 67
Competition engines....................... 70
Camshaft timing 71
Camshaft summary 72

Chapter 8. Valve springs 73
Valve spring dimensions 74
Standard valve spring data............. 75
Measuring valve spring poundage.. 77
Advertised valve lift........................ 79
Competition engines....................... 80
Valve spring summary.................... 80

Chapter 9. Rockers & rocker geometry 81
Rocker geometry criteria 81
Valve stem height........................... 82
Altered rocker geometry 82
Lash caps 83
Rocker sizes/designs..................... 83
Checking rocker geometry 86

Chapter 10. Exhaust systems......... 89
Standard cast iron exhaust manifold...................................... 89
Exhaust system construction 89
Four into two into one 90
Four into one.................................. 90
Primary pipes................................. 90

Chapter 11. Flywheel & clutch. Engine balance 93
Flywheel... 93
Engine balance 95

Chapter 12. Ignition system........... 96
Distributor spindle.......................... 97
Drive gear....................................... 97
Endfloat/endplay............................ 97
Contact breaker points 97
Condenser 98
Electronic module.......................... 98
Distributor cap 98
Rotor arm....................................... 99
Coil... 99
High tension wires 100
Sparkplugs................................... 100
Checking spark quality 100
Ignition timing marks 102
Static advance............................. 104
Total advance............................... 105
Vacuum advance 105
Ignition timing setting and checking 105
Rev-limiters.................................. 105
Ignition system summary............. 105

Chapter 13. Carburettors............. 107
Throttle action.............................. 109
Carburettor summary................... 109
Inlet manifolds 109
Air filters....................................... 110
Ram pipes.................................... 110
Fuel supply 110

Chapter 14. Sierra Cosworth & Cosworth-headed Pinto engines.. 111
Introduction.................................. 111
Compression ratio (cr) – Pinto & Cosworth blocks 113
Cosworth head/Pinto block camshaft drive modifications 114
Cylinder head porting 114
Camshafts.................................... 116
Valves... 119
Valve springs................................ 120
Camshaft followers 124
Cylinder head rebuild................... 125
Short block 128
Pistons... 128
Piston rings.................................. 129
Valve clearance............................ 129
Crankshaft 130
Connecting rods 130
Piston pin oiling modifications..... 131
Fitting cylinder head 132
Camshaft timing 134
Valve to piston clearance – checking 135
Ignition system............................. 136
Exhaust system 137
Carburettors................................. 138

Chapter 15. Starting engines & oiling requirements 141

Index 143

Veloce SpeedPro books –

978-1-903706-59-6 978-1-903706-75-6 978-1-903706-76-3 978-1-903706-99-2 978-1-845840-21-1 978-1-787111-68-4 978-1-787110-01-4

978-1-787111-69-1 978-1-787111-73-8 978-1-845841-87-4 978-1-845842-07-9 978-1-845842-08-6 978-1-845842-62-8 978-1-901295-26-9

978-1-845842-89-5 978-1-845842-97-0 978-1-845843-15-1 978-1-845843-55-7 978-1-845844-33-2 978-1-845844-38-7 978-1-787113-34-3

978-1-845844-83-7 978-1-787113-41-1 978-1-845848-33-0 978-1-787111-76-9 978-1-845848-69-9 978-1-845849-60-3 978-1-787110-91-5

     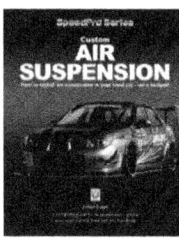 

978-1-845840-19-8 978-1-787110-92-2 978-1-787110-47-2 978-1-903706-94-7 978-1-787110-87-8 978-1-787111-79-0 978-1-787110-88-5

978-1-903706-78-7 978-1-787113-18-3 978-1-787112-83-4

– more on the way!

Introduction, using this book & essential information

INTRODUCTION

The aim of this book is to tell you how to modify the single overhead camshaft (SOHC) Ford 'Pinto', Cosworth-headed Pinto or Sierra Cosworth 'Pinto' engine and end up with an engine that is reasonable powerful and reliable. Whether you want extra power for road or competition use, you'll find the information you need within this book. Although this book specifically deals with naturally aspirated (carburettors, no compressor) engines, because the broad aims of tuning for high performance (minimum friction, maximum gas flow, maximum efficiency and reliability) are the same, most of the information is also applicable to turbocharged/supercharged engines and fuel-injected engines. If you are intending to use forced induction, you'll need to talk to the equipment supplier about optimum compression ratios, rpm requirements and any other aspects of engine specification which will be affected by your choice of equipment.

The Ford single overhead camshaft (SOHC) 'Pinto' engine was made in vast numbers to power a great many different vehicles and, although no longer in production for passenger cars, it is still in wide use. This all cast iron engine is quite large and very heavy, but it's also readily available, spares are reasonably priced, there are plenty of tuning components available and a very good amount of extra power is attainable at relatively low cost.

All three engines in the Pinto series covered by this book share the same crankshaft stroke (76.95mm/3.029in). The bore sizes of the three capacities are 81.32mm/3.201in (1600cc), 86.20mm/3.394in (1800cc) and 90.82mm/3.575in (2000cc). There are other Pinto engines which this book does *not* cover, namely the 1300cc engine (bore and stroke of 79.02mm by 66.0mm) the 1600cc 'E-Max' engine (bore and stroke of 87.67mm by 66.0mm) and the American-built 2300cc Pinto engine which, while similar in design to the European units is, in fact, different in almost every detail.

Pinto engines have several weak points which place some limitation on how far they can be taken in terms of extra power. However, the engine's evolution has resulted in improvements which made the unit more reliable than when first introduced. Specialist high-performance manufacturers have also designed and developed components which have contributed to the removal of virtually all of the problems

SPEEDPRO SERIES

associated with this unit, meaning the engine can now be modified for high-performance use with real confidence.

These weak areas are the connecting rods, the camshaft lobes and rockers, the centre and rear camshaft pillars and the camshaft centre pillar bearing. Details of the problems and what to do about them are covered in chapter 1.

There is now a huge amount of equipment available for SOHC Pinto engines and naturally aspirated Sierra Cosworth engines. Before embarking on building any level of engine you should arm yourself with as much information as possible with regard to suppliers of parts and the latest trends in technology. In spite of the fact these are old obsolete engines there is still a little bit of development going on.

There are several firms in England who make excellent componentry for these engines. Some of them are:

Vulcan Engineering
www.vulcanengines.com
Tel: +44 1474 874 689

Burton Power
www.burtonpower.com
e-mail: sales@burtonpower.com
Tel: +44 20 8518 9136
fax +44 20 8554 4828

Holbay of Grundisburgh
www.holbay.co.uk
e-mail: richard@holbay.co.uk
Tel: +44 1473 738 738
Fax: +44 1475 738 739

Millington Engineering
Tel: +44 1476 738 738
Fax: +44 1476 789 692

All of these companies make good gear and have comprehensive catalogues which can be looked at on the internet and copied. The Burton Power printed catalogue is extremely comprehensive, they and the other businesses will send all relevant details by post on request. These companies will send components and engines to anywhere in the world.

Sierra Cosworth engine

This book also deals with the double overhead cam (DOHC) four valve per cylinder Sierra Cosworth engine, which is essentially a derivative of the Pinto unit. Although primarily built for use with a turbocharger, more and more of these engines are being converted to natural aspiration – and with very good results. The Cosworth cylinder heads are reasonably plentiful and can be fitted to a Pinto block: Pinto engines so equipped produce excellent power. Nearly 40,000 Sierra Cosworth engines were built.

USING THIS BOOK

Throughout this book the text assumes that you, or your contractor, will have a workshop manual specific to your engine to follow for complete detail on dismantling, reassembly, adjustment procedure, clearances, torque figures, etc. This book's default is the standard manufacturer's specification for your model so, if a procedure is not described, a measurement not given, a torque figure ignored, you can assume that the standard manufacturer's procedure or specification for your engine needs to be used.

You'll find it helpful to read the whole book (whether your engine is Pinto, Sierra Cosworth or a Cosworth-headed Pinto) before you start work or give instructions to your contractor. This is because a modification or change in specification in one area will cause the need for changes in other areas. Get the whole picture so that you can finalize specification and component requirements as far as is possible before any work begins.

For those wishing to have even more information on high-performance short block building principles, ignition systems and Weber or Dellorto sidedraught carburettors, the following Veloce titles are recommended further reading: *How To Blueprint & Build a 4-Cylinder Short Block For High Performance*; *How To Build & Power Tune Distributor-type Ignition Systems*; *How To Build & Power Tune Weber & Dellorto DCOE & DHLA Carburetors*; *How To Choose Camshafts and Time Them for Maximum Power*.

ESSENTIAL INFORMATION

This book contains information on practical procedures; however, this information is intended only for those with the qualifications, experience, tools and facilities to carry out the work in safety and with appropriately high levels of skill. Whenever working on a car or component, remember that your personal safety must *always* be your *first* consideration.

The publisher, author, editors and retailer of this book cannot accept any responsibility for personal injury or mechanical damage which results from using this book, even if caused by errors or omissions in the information given. If this disclaimer is unacceptable to you, please return the pristine book to your retailer who will refund the purchase price.

In the text of this book **"Warning!"** means that a procedure could cause personal injury and **"Caution!"** that there is danger of mechanical damage if appropriate care is not taken. However, be aware that we cannot foresee every possibility of danger in every circumstance.

INTRODUCTION, USING THIS BOOK & ESSENTIAL INFORMATION

Please note that changing component specification by modification is likely to void warranties and also to absolve manufacturers from any responsibility in the event of component failure and the consequences of such failure.

Increasing the engine's power will place additional stress on engine components and on the car's complete driveline: this may reduce service life and increase the frequency of break- down. An increase in engine power, and therefore the vehicle's performance, will mean that your vehicle's braking and suspension systems will need to be kept in perfect condition and uprated as appropriate. It is also usually necessary to inform the vehicle's insurers of any changes to the vehicle's specification.

The importance of cleaning a component thoroughly before working on it cannot be overstressed. Always keep your working area and tools as clean as possible. Whatever specialist cleaning fluid or other chemicals you use, be sure to follow – completely – manufacturer's instructions and if you are using petrol (gasoline) or paraffin (kerosene) to clean parts, take every precaution necessary to protect your body and to avoid all risk of fire.

Chapter 1
Problem areas

STANDARD CONNECTING RODS
Forged connecting rods

In all Pinto engines the standard forged connecting rods are reliable to about 6700rpm with the standard weight pistons and piston pins fitted but are prone to breakage if the engine is subjected to sustained or continuous revs over 6700rpm. The point of breakage is almost always about 25mm/1in below the piston pin (gudgeon pin) boss.

The connecting rod is basically well-designed in most areas. The big end is well-proportioned with 9mm diameter bolts. The small end (little end) of the connecting rod is quite large and, if anything, over-built but the all important I-beam of the connecting rod is marginal, considering the weight of the piston and piston pin. To be fair, this is a road-going production engine and not a racing engine. For all normal use these connecting rods are more than adequate, and connecting rod failure has never been a problem on standard engines used normally.

From a high-performance point of view, the standard forged connecting rods don't 'look right' and, once these engines are tuned to produce more power and run to higher revs, the connecting rods break regularly. True, the connecting rods will turn high rpm (8000rpm and more) for a very short length of time, but this is hardly the reliability needed for high performance applications. Constantly changing connecting rods to preclude failure – and even then not really knowing whether or not the rods are going to break – is not normal race engine practice.

The reliability of this connecting rod is improved by approximately 300 rpm (maximum rpm 7000) if the small end boss is lightened and a lightweight forged piston and piston pin used. The forging flash on the I-beams is not removed or touched in any way so that the maximum amount of material is kept. Although polishing the sides of the I-beam would be desirable, there's just not the material available to remove anything without detriment to the strength of the rod (the lesser evil).

Later, cast connecting rods

The later 'wide' cast connecting rod, as found in all 2000IS engines and all 1988 on Sierra and Transit engines with a large '205' cast in the block, is a definite improvement over the original rod, but it, too, has a strength limitation in the I-beam. If the early forged connecting rod was of the same proportions as the cast version there would be few breakage problems.

Regard the standard cast connecting rods as useable for applications where up to 6900rpm is required on a more or less continuous basis with standard pistons and piston pins, and 7200rpm when lightweight forged pistons and piston pins are fitted. Anything above these engine speeds has an element of risk attached and standard connecting rods subjected to higher rpm, such as 7500rpm, must be changed frequently to prevent breakage. Consider a normal season's racing (800 kilometres/500 miles) as the life of one set of these standard connecting rods.

PROBLEM AREAS

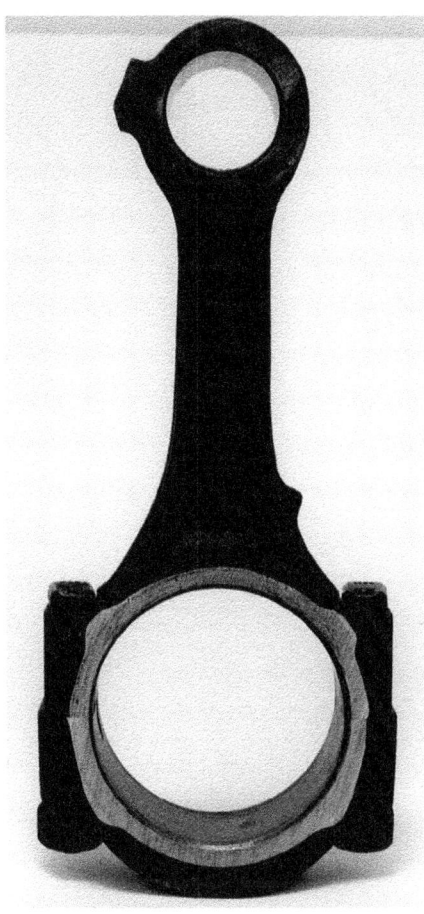

Early forged standard connecting rod.

Forged standard connecting rod with lightened small end.

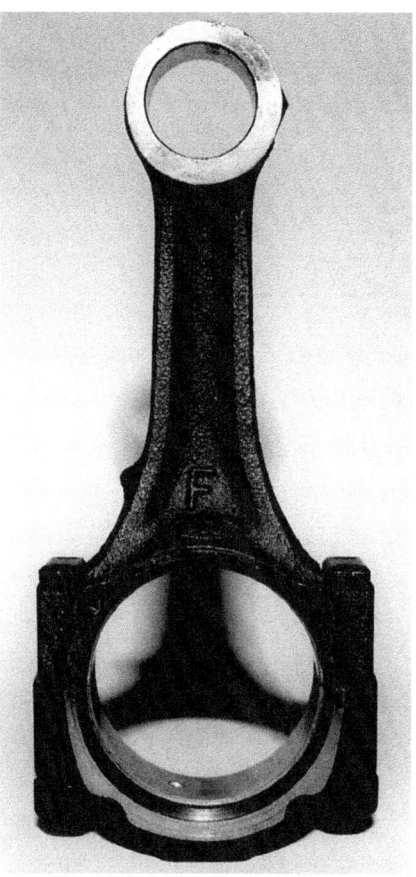

Later cast steel standard connecting rod is capable of 7500rpm with a lightweight piston and piston pin and 7200rpm with the standard piston and piston pin.

The later cast connecting rod should be used just as it is (*no polishing*). Ideally, you should use brand new connecting rods which have been straightness tested and crack tested. Do not use any standard-type connecting rods that have previously been used in a competition engine. Next to new rods, used rods out of passenger car engines that have never been stressed by high rpm usage (5000rpm and above) are best.

Cosworth rods

The Sierra Cosworth connecting rod is very strong (as in being bullet proof to 9000rpm) and is a standard Ford part which is a near 'drop in' fit for Pinto engines and it does solve the problem of Pinto connecting rod failure, but these rods are a comparatively recent arrival. Note that using the Sierra Cosworth connecting rod in conjunction with a standard type Pinto piston will require the use of an alternative piston pin retention method (such as Teflon buttons) because the small end is of the floating type on the Sierra Cosworth and not a press fit like the Pinto's.

Caution! The Sierra Cosworth connecting rod is 1.5mm/0.060in longer than the Pinto rod and this means that, depending on the piston used, it may be necessary to machine the top of the piston to compensate (Cosworth pistons have a lower piston pin to crown height). Holbay Engineering can supply custom machined forged pistons for this

Standard distributor rotor on the left and governor rotor on right.

SPEEDPRO SERIES

combination which will see the top of the piston flush with the top of the block at top dead centre (TDC).

Standard connecting rods - summary

The standard early forged connecting rod with a standard piston and piston pin fitted has a continuous rpm rating of 6700rpm. These rods (with standard pistons and piston pins fitted) will break above 6700rpm after some use. The reliability of the connecting rod is improved if the small end of the connecting rod is lightened to the extent that the wall thickness of the small end is reduced to 3.5mm-4.0mm/0.138in-0.157in and further improved if lightweight pistons and piston pins are fitted.

The standard piston, piston pin and rings together weigh 767gm/27.05oz. The weight reduction brought about by changing to lightweight pistons, piston pin and rings that together weigh about 515gm/18.16oz, plus the reduction of about 65 grams created by removing material from the small end of the connecting rod, reduces the overall weight acting on the I-beam of the connecting rod by around 580gm/20.45oz and lifts the maximum rpm rating to about 7000rpm.

The later cast connecting rods are slightly stronger, but not much. Changing the piston, piston pin and piston ring assembly to one that weighs 515gm/18.16oz instead of the usual 765gm/26.98oz improves the reliability of the connecting rod to an absolute maximum of 7500rpm but *not* for applications where these revs are used on a continuous basis.

If the standard connecting rods are going to be retained, the fitting of the lightest possible piston, piston pin and ring combination is recommended. The cast standard connecting rod weighs 680 grams, while the forged standard connecting rod weighs 700 grams (or 635 grams lightened as suggested).

Nothing can alter the fact that the two standard type connecting rods are not suitable for all-out competition use and, as a consequence, alternative connecting rods have to be used. With the lightest forged pistons fitted to either of the standard connecting rods, the maximum revs possible are 7300-7500rpm with limited reliability. The connecting rods will only take this sort of treatment for a limited period of time and it only takes one connecting rod to break and the engine will be totally wrecked ...

With either type of standard connecting rod fitted, maximum engine rpm *must* be limited to avoid connecting rod failure. Use a governor-type distributor rotor (readily available for the standard Bosch distributor used on many of these engines). Governor rotor cut-out speeds start at 6200rpm and the cut-out rpm is cast on the rotor. Alternatively, an electronic rpm limiting device can be fitted into the ignition system. The use of both methods will give peace of mind.

The two weak camshaft pillars.

CAMSHAFT LOBES AND ROCKERS

Many standard Pinto engines have had camshaft failures of one sort or another. The oil spray bar is usually blamed (it can be a source of problems if an oil hole becomes blocked) but, in reality, there is so much oil from all the rockers flying around that this idea can usually be discounted. The real problem on standard engines was one of rocker hardness and camshaft lobe hardness.

The original rocker geometry of the standard engine was always correct, but it certainly wasn't after a replacement camshaft with a different base circle diameter (any significant amount – 1.0mm/0.040in, plus) was fitted. On high-performance engines this is where the real problems started because this was a new factor unrelated to the original rocker/cam lobe surface hardness problem.

Early standard engines often had the problem of one or two rockers (or more) and, perhaps, the cam lobes wearing away rapidly. On checking surface hardness of the worn rockers it was common to find the hardness value slightly down on that of the surviving rockers, even if the surviving

PROBLEM AREAS

Auxiliary shaft and the gear which can wear.

rockers looked to be on the point of failing themselves but were actually still giving good service. The tops of the camshaft lobes would also show around 0.75mm/0.030in wear even though the engine would still be running well, if noisily.

There are now plenty of camshaft manufacturers (including Ford) making complete kits for these engines. Because of the known problems, replacement camshafts and rockers are all checked for sufficient hardness. Outright failures are few and far between, although the overall wear characteristics remain unchanged.

CAMSHAFT PILLARS

The front pillar is extremely strong and never causes any problem. The centre and rear pillars are extremely weak and do not represent good design. These two items must always be handled with extreme care to avoid damage (breaking them off). Ford never saw fit to improve the strength of the pillars during the life of the Pinto engine.

The material thickness of these pillars is marginal at best and, further to this, the factory drills an oil feed hole for the spray bar in the middle of the centre pillar – at the thinnest point on one side!

The centre and rear pillar can be strengthened to a satisfactory level, but this involves detailed engineering work and the brazing of mild steel straps over them. These modified pillars will not break even in the most rigorous of service. A close fitting steel mandrel (0.0005inch/0.013mm) has to be made that fits into the centre and rear pillars tunnels (bearings out) and then the pillars have the straps braized on. The mandrel prevents distortion occurring during braizing. As a further alteration the oil feed to the rockers is not taken off the centre pillar but rather off the front and rear pillars only. This means making up a new spray bar and drilling into the pillars (involves some re-work). What this does is allow the oil fed to the highly stressed bearing in the centre pillar to oil the bearing only and not be drained off to feed the rockers as well. If the valve spring pressure and the lift are kept within reasonable limits and the geometry is correct the standard pillars do not normally break in a high-performance application.

CAMSHAFT BEARINGS

The early centre and rear camshaft bearings (white metal type) would also wear out prematurely on standard engines. On early engines fitted with white metal camshaft bearings, the centre bearing would invariably be well worn after even a moderate (50,000km/30,000 mile) usage. Later standard engines feature hard wearing bronze bearings and, while the underlying problem is not actually resolved by this modification, the symptoms are reduced to an acceptable level.

The centre bearing takes the maximum flex from the camshaft (caused by the valve spring and camshaft action) and this is why it suffers first from wear problems. Centre bearing wear is increased when strong valve springs and a high lift camshaft are installed.

Just to exacerbate the problems caused by a worn centre bearing, the oil spray bar is fed from this bearing.

PISTON TO VALVE CONTACT

2000cc engines have deeper combustion chambers and so, when fitted with a standard camshaft, do not suffer piston to valve contact even when a cam drivebelt breaks. On the 1600cc and 1800cc engines (with standard camshaft) if the drivebelt breaks valves will be bent.

Any Pinto engine can have inlet and exhaust valve reliefs professionally machined into the tops of the pistons to prevent piston to valve contact – this is particularly important for road cars where reliability is essential. If the cylinder head is planed a lot, and the camshaft has more lift than standard, machining deep enough valve reliefs becomes difficult (regard 3mm as a safe maximum) but whatever valve relief depth can be safely obtained should be obtained. Reliability is the most important attribute of any high-performance engine.

AUXILIARY SHAFT GEAR WEAR

When assembling the short block the first to be checked is the mesh of the distributor drive gear with the auxiliary shaft gear. If an auxiliary shaft is found to have its gear teeth worn to a knife-edge, the reason for this is poor gear mesh. This is not a common problem, but it does crop up occasionally and will cause grossly fluctuating ignition timing.

The procedure for correction (using good used parts or brand new parts) is to install the auxiliary shaft and location plate, then rotate the shaft by hand to check for freedom of rotation and also check the endfloat (lash). Endfloat should be kept to 0.125mm/0.005in.

The next step is to fit the distributor you are going to use, oil drive shaft and oil pump. Once they are all bolted in, rotate the auxiliary shaft clockwise and then anti-clockwise, note if there is any difference in the effort required to turn the shaft in different directions or whether rotating one way feels a bit 'gritty'- this 'grittiness' is easily felt when turning the auxiliary shaft by hand via the belt drive cog. Correctly matched gears have very low drag in both directions. If you feel that all is not well, remove the location plate and make up a packing piece 0.5mm/0.020in thick and place this behind the location plate. This packing will shift the auxiliary shaft forward, and there should be a resultant reduction in drag as mesh is improved. It's possible that more than 0.5mm/0.020in will be necessary (up to 0.75mm/0.030in). The repositioned auxiliary shaft will not normally suffer gear wear again.

LOOSE SPROCKETS

Check the fit of the drivebelt sprockets on the crankshaft, auxiliary shaft and camshaft. If the fit of any sprocket is loose (as opposed to a tap on fit) it will eventually become very loose and will rattle when the engine is running. If a securing bolt for any one of these three sprockets is left loose, or works loose, the engine will usually emit a knocking noise. If left too long in this loose condition the parts concerned will be seriously damaged and require replacement.

VALVE GUIDES

The standard valve guides are integral cast iron ones and they do wear. The best method of restoring standard worn valve guide bores to better than original is to have K-Line inserts fitted to them. Many engine machine shops/engine reconditioners have this equipment.

www.velocebooks.com / www.veloce.co.uk
All current books • New book news • Special offers • Gift vouchers

Chapter 2
Short block components

PISTONS AND CONNECTING RODS

The choice of parts depends on the application. The majority of modified engines (1600, 1800 and 2000) are built using standard parts such as oversized standard cast pistons and connecting rods in freshly rebored cylinders. There is little point in using a block which has more than 0.05mm/0.002in bore wear: in fact, there's very little point in modifying an engine that has any bore wear at all. There is *no* substitute for a perfectly parallel cylinder bore. AE cast aluminium over size pistons are available for all Pinto engines in plus 0.020inch/0.5mm, 0.030inch/0.75mm, 0.040inch/1.0mm and 0.060inch/1.5mm over sizes.

The standard cast aluminium pistons for all Pinto engines are rated as being suitable for use up to about 7000rpm; which means that the standard pistons are slightly stronger (in rpm terms) than the standard connecting rods. An engine used on the race track equipped with standard

Standard 2000cc Pinto piston and pin.

type cast pistons and standard connecting rods and being revved consistently to 7000rpm will most likely end up having a connecting rod failure rather than a piston failure.

In most instances it is the 2000cc version of the Pinto engine that gets modified because it is the largest capacity engine. Most of the alternative equipment is made to suit these engines so the following information refers mainly to them.

A standard type cast piston is available from AE in plus 0.090inch/2.25mm for 2000cc engines which takes the capacity out to 2.1 litres. There is no weight difference over the standard piston. These pistons are 'drop in fit' items once the block has been bored out, and are compatible with the standard type connecting rods.

SPEEDPRO SERIES

Note that engines (with components based on standard units) that survive occasional high rpm use (7000rpm plus) do so because the components are not continuously subjected to this sort of treatment. Occasional short duration revving to 7500rpm is not the same, in component stress terms, as revving to 7500rpm at each and every gear change.

Lightweight forged pistons are available from specialist piston manufacturers such as Omega, Accralite, Mahle and Holbay and will withstand 9000rpm plus. These pistons all have 24mm diameter piston pins and round wire circlips. When using standard type Pinto connecting rods in conjunction with these forged pistons (not to 9000rpm), such pistons will allow the standard method of piston pin retention (interference fit) in the connecting rod to be used or, alternatively, honing out the rod's piston pin tunnel to give a 0.010mm/0.0004in clearance and a fully floating piston pin.

To reduce 'ring drag' these lightweight forged pistons have rings of narrow section – top 1.0mm/0.040in; second 1.5mm/0.060in; oil control 3.0mm/0.118in.

Caution! *Never* re-use round wire circlips.

New standard connecting rods are going to last longer than used standard connecting rods (all things being equal) but continually fitting new sets of standard connecting rods is false economy unless competition class rules require original equipment parts to be used. Buying non-standard heavy duty 'bullet proof' connecting rods in the first place is the most cost effective method for engines that will be regularly required to rev at over 7000rpm.

Kolbenschmidt Ford V6 2.8-litre 93mm piston.

Buying expensive non-standard heavy duty 'bullet proof' connecting rods as made by Holbay, Arrow or Farndon, for example, in the very first instance is cost effective in the long run as the connecting rod failure problem essentially ceases to exist. Farndon, for example, make three lengths of connecting rod for Burton Power. These connecting rods' centre to centre lengths are standard at 5.000inch/127.0mm, 5.060inch/128.5mm (which is the Sierra Cosworth centre to centre distance) and 5.150inch/130.8mm.

The 5.150inch/130.8mm centre to centre distance connecting rods are designed to fit into engines with specially machined Accralite 91mm, 92mm and 93mm forged pistons which have gudgeon pin holes placed higher in the piston than standard. Accralite, as a consequence, make two very similar pistons of the same diameter sizes but they do NOT interchange. This piston and connecting rod combination is designed to reduce the connecting rod angulation to the minimum possible within the confines of the engine design. Holbay do the same basic combination (long connecting rod, short piston height) to order. This is the ideal setup for these engines even though it is expensive.

An alternative is to fit connecting rods out of a 1600cc Fiesta diesel engine. These connecting rods are available at a very reasonable price new, and even cheaper second-hand from a scrapped engine. The other advantages of using the Fiesta diesel engine connecting rods is that they are very strong and will withstand 8000rpm on a continuous basis when the small end is lightened; their longer centre to centre dimension (0.125in/3.2mm) also reduces the connecting rod angulation.

There is, however, some complication with using Fiesta rods as they are not a 'drop in' fit. The small ends of the 1600cc diesel connecting rod are of the floating type and so Teflon buttons will have to be used to locate the piston pin in a Pinto application as the standard type press fit piston is going to be used (that's genuine Ford V6 pistons, AE replacement Ford V6 pistons or Kolbenshmidt V6 replacement pistons). The problem with these connecting rods is that the big end

SHORT BLOCK COMPONENTS

1600cc Ford diesel connecting rod (small end has been lightened).

Kolbenschmidt piston and Fiesta 1600 diesel connecting rod assembly, note the Cosworth connecting rod bolts (a good move).

bearing tunnel diameter is much smaller than the Pinto's (in fact, 3.0mm/0.118in smaller). The Pinto crankshaft's big end journals can be ground down undersize from the standard 52.00mm to 49.00mm and the journals widened slightly to suit the diesel connecting rod but this does weaken the crankshaft slightly, on a big end journal cross sectional area basis, but the crankshafts do not have a tendency to break. A maximum sized 'fillet radius' can and should be ground into the corners of each big end journal when this conversion is done. The larger the journal corner radius the stronger the crankshaft! Tests have proved that the combination is safe for 7500rpm use with many users turning their now 2090cc engines to 8000rpm and are not having failures.

When the crankshaft is reground to take the Fiesta diesel connecting rods, the crankshaft's big end bearing journals can also be 'offset ground' which will increase the stroke slightly. If there is no cc restriction this move results in an engine that has more torque. The crankshaft can be 'offset ground' and the stroke increased by approximately 2.8mm which means that the stroke of the engine will be about 79.8mm. This bore and stroke combination results in 2168cc. The revs must not exceed 7500 if bottom end reliability is to be maintained. The piston crowns will also protrude well above the top of the block and will have to be machined to avoid contact with the cylinder head.

The Fiesta diesel connecting rods can be ground down on their sides to fit the standard crankshaft's journal width, instead of altering the crankshaft journal width to suit the connecting rod. The Fiesta diesel connecting rods are 1.058in/26.95mm thick whereas the Pinto connecting rod is 1.018in/24.9mm thick. This is a relatively easy machining operation which is done on a surface grinder.

The Kolbenschmidt piston weighs 700gm/24.69oz (including the rings and piston pin) while the comparable standard Pinto piston, piston pin and rings weigh 767gm/27.05oz. A lightened 1600cc diesel connecting rod will weigh approximately 700gm/24.69oz which is similar to the standard connecting rod.

The advantage of using the diesel connecting rods and the Kolbenschmidt pistons is the strength factor of the pistons and the diesel connecting rods, the reduction in connecting rod angulation (via the slightly longer connecting rod) gives smoother high rpm running, the increase in engine capacity and cost-effectiveness. The piston crown will also be slightly above the surface of the block at TDC. This mix of components requires reworking some of the components but, once completed, offers great strength and reliability at a relatively small cost. Note that second-hand Fiesta diesel rods are fine (they're never stressed in their original application) if checked thoroughly and, even if you buy them

SPEEDPRO SERIES

Holbay TT connecting rod weighs 572 grams without the bearing shells.

This piston is made for Holbay by Accralite for one of their crankshaft and connecting rod combinations. This piston weighs 426 grams and the gudgeon pin weighs 93 grams. The gudgeon pin is 20.64mm/0.8125inch in diameter and designed to suit both types of Holbay connecting rods.

Holbay's 'H' section forged connecting rod.

new, the cost will be around one third that of aftermarket high-strength rods. The fitting of Kolbenschmidt pistons, Fiesta diesel connecting rods, and 'offset grinding' the crankshaft involves a considerable amount of effort and cannot be considered an easy option.

When it is decided to change the standard connecting rods and pistons to improve the reliability of the bottom end of the engine, by far the easiest way is to buy 'drop in fit' pistons and connecting rods, even if they do cost considerably more than any other option. This is definitely the recommended way to end up with a 'bullet proof' engine.

Using Fiesta diesel connecting rods and Kolbenshcmidt pistons are mentioned in this book because it is an option. It involves a considerable amount of re-working of the engine componentry, though, and, as a consequence, it is not recommended that anyone go down this route unless you really know what you are in for.

Holbay make two types of 'bullet proof' connecting rods and both are forged out of EN24 steel. The original Holbay connecting rod was the 'H' section one while the later alternative connecting rod is the TT one. Both connecting rods are available in two widths to suit the standard crankshaft big end journal width size or Holbay's narrow big end journal width size made to suit the billet crankshafts (that's standard at 26.0mm/1.125inch or narrow at 23.8mm/0.938inch). The bearing shell inserts are the same in either width connecting rod.

These two connecting rods are available in the standard 5.000inch/127.00mm centre to centre length and 5.241inch/133.12mm but up to 5.514inch/140.00mm centre to centre distance is available. The small ends are bushed and the usual diameter is 0.8125inch/20.64mm gudgeon pins, as found in Holbay's

SHORT BLOCK COMPONENTS

forged pistons (standard size for interference fit gudgeon pin retention is available).

Piston & connecting rod weights

Standard piston – 565gm/19.92oz.
Standard ring set – 48gm/1.69oz.
Standard piston pins weigh 154gm/5.43oz.
Standard piston, pin and rings – 767gm/27.05oz.
Lightweight forged piston, pin and rings weigh 510gm/17.98oz.
Kolbenshmidt piston pin and rings – 705gm/24.86oz.
Standard 1600 diesel connecting rod – 800gm/28.21oz.
Modified and lightened 1600 diesel connecting rod – 700gm/24.69oz.
Standard forged connecting rod – 700gm/24.69oz.
Lightened standard connecting rod – 635gm/22.39oz.
Later cast connecting rod – 680gm/23.98oz.
Sierra Cosworth connecting rod – 720gm/25.39oz.
Aftermarket connecting rods on average are – 525gm/18.53oz.

Piston & connecting rods - summary

The standard pistons of all three engines (1600, 1800 and 2000) are capable of reliable operation to 7000rpm.

The standard forged connecting rod is capable of withstanding 6700rpm on a continuous basis with a standard piston.

The later standard cast connecting rod is capable of withstanding 7000rpm (just) on a continuous basis with a standard piston.

The later cast connecting rod fitted with a lightweight forged piston is capable of withstanding 7500rpm, but not on a continuous basis. This

Machined hole which can be used for a dowel.

The special spotting punch made to mark the back of the flywheel – the punch must be hardened.

means changing the connecting rods every racing season or every 300 racing miles/500 kilometres.

The fitting of lightweight forged pistons to either of the standard connecting rods does improve the reliability of the connecting rods but not by much (300rpm).

Kolbenshmidt 93mm cast pistons and Fiesta 1600 diesel rods will withstand 8000rpm. The crankshaft and connecting rods have to be suitably modified but the combination is strong and offers 2.1-litre capacity.

For racing purposes (any continuous operation above 6700rpm) where reliability is essential the connecting rods and pistons need to be changed for stronger items (forged pistons and alternative connecting rods).

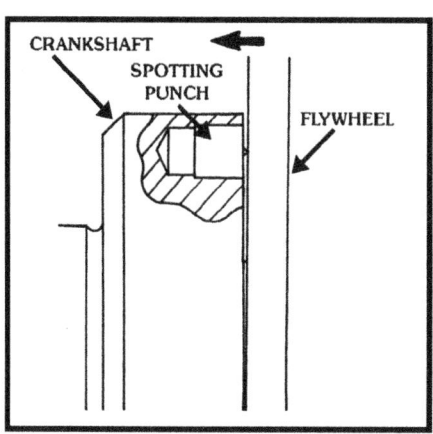

How the spotting punch is used.

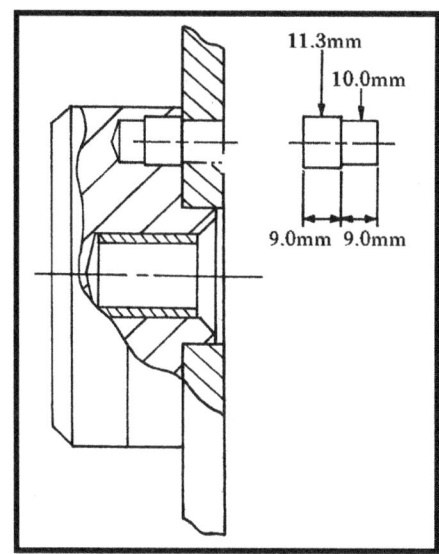

Dowel dimensions and fitted location.

For pure racing engines keep the bore as near to the standard size as possible and definitely within 1.5mm/0.060in oversize.

The standard Sierra Cosworth connecting rod is near 'bullet proof' when installed in a Pinto and would be a 'drop in' fit if it wasn't for the piston pin retention situation and the fact that these rods are 1.5mm/0.060in longer, which will almost always mean that the piston tops will need to be machined to give sufficient head clearance. There are many alternative

SPEEDPRO SERIES

connecting rods available which are of excellent quality and do not break. Some aftermarket connecting rods are quite light in weight while still offering excellent strength and reliability.

The use of Sierra Cosworth connecting rods and lightweight flat-topped forged pistons, which have round wire circlips to facilitate piston pin retention, results in a near unbreakable combination. This combination is relatively inexpensive considering the quality of the componentry and absolute reliability available at very high rpm (connecting rods and pistons will withstand 9000rpm).

Note that Burton Power stock Accralite forged pistons which have semi-finished tops which can be custom machined to suit any known application. These semi-finished pistons are available in 91mm (90.93mm actually), 92mm and 93mm diameter sizes. The 90.93mm diameter size is for classes of racing which do not allow the swept volume to exceed 2000cc. All Accralite Pinto pistons have a 1.0mm thick top compression ring, a 1.2mm thick second compression ring and a 2.7mm thick oil control ring.

Although valve reliefs can be professionally machined in any piston crown, the majority of modified Pintos do not have this feature: instead, it's accepted that if the cam drivebelt breaks, valves will be bent.

CRANKSHAFT AND FLYWHEEL

The standard crankshaft will take 8500rpm plus, and does not need to be replaced with a stronger forged item. The forged crankshaft from a Sierra Cosworth can be directly substituted for the Pinto's but, although it is definitely stronger, it's heavier too (increasing engine inertia). It also has nine flywheel retention bolts and a different flywheel for the large Sierra Cosworth clutch. The fitting of a Sierra Cosworth crankshaft, or an aftermarket forged item, is not deemed necessary because a modified standard Pinto cylinder head is not efficient enough to make power over 7500rpm, although in racing, engines always get revved more than the point of maximum power.

With the standard crankshaft limit known to be 8500rpm, the fitting of lightweight aftermarket connecting rods and lightweight forged pistons is clearly required to ensure reliability up to this rpm point. Fitting the longer, 5.150inch/130.8mm, centre to centre distance connecting rods, and a shorter gudgeon pin centre to piston crown height, is the unbeatable scenario.

In high-performance applications, standard-based Pinto crankshafts do not break – although the flywheel bolts have been known to break or come loose. All road-going engines should have the threads of the six bolts coated with a locking agent such as Loctite to prevent the bolts from undoing. Always fit new standard flywheel bolts or, preferably, uprated (higher strength) bolts.

Competition engines require the flywheel to be dowelled to the crankshaft by, at the very least, one large dowel. The crankshaft already has a drilled hole in the flywheel flange and the centre of this can be 'spotted back' onto the flywheel and a pilot hole, followed by a 10.0mm/0.393in hole, drilled through the flywheel. A special dummy spotting punch has to be made up to 'spot' the centre of the existing hole in the crankshaft onto the flywheel.

The point in the centre of the spotting punch must be slightly proud of the flange face when the punch is in place in the hole. The flywheel is then bolted to the crankshaft which will leave a centre punch mark on the flywheel. The centre of this punch mark is picked up using a combination centre drill (which will not wander) and then a normal twist drill bit is used to bore the hole right through: the drilled hole is then reamed to ensure perfect sizing. The size of the hole in the crankshaft flange is 11.2mm/0.437in so a special stepped dowel will have to be made up to suit the specific application – the step will keep the dowel in place. The dowel *must* be a tight fit in the crankshaft.

The idea of dowel/s is to protect the flywheel securing bolts from the crankshaft's torque. If the flywheel has bolts only, they must clamp the flywheel to the crankshaft and take twisting torque.

The main reason for utilizing the factory drilled hole in the crankshaft flange is the difficulty of accurately drilling new holes into the flange. If more than one dowel is to be fitted, the crankshaft has to be vertically mounted on to a radial drill or horizontally mounted on a horizontal boring machine. This way holes can be accurately drilled and reamed into the crankshaft and flywheel. This sort of work can be carried out by an engineering machine shop: *not* an engine machine shop. The holes are generally drilled through the flywheel and into the crankshaft and then reamed to be slightly undersize on a 8mm/5/16in dowel size to effect an interference fit.

The crankshaft and flywheel are removed from the machine and the dowels firmly tapped into the crankshaft. Next, the flywheel is reamed out with an 'on size' reamer which will give the dowel/s clearance in the flywheel holes (0.02mm/0.0003in). The protruding

SHORT BLOCK COMPONENTS

Indestructible Holbay solid billet crankshafts are all cross drilled in the journals (that's two standard holes feeding oil to each bearing) and made out of EN40B material. They are available in standard stroke 77.1mm, 88mm and 90mm. Holbay will make any crankshaft to order up to 90mm. The 88mm one is the most popular one.

This top quality Holbay aluminium flywheel weighs 3.28 kilograms/7 pounds 4 ounces and is supplied balanced. This flywheel has six bolt retention and three dowel holes in it.

crankshaft. Their 2239cc engine has a 90.0mm/3.550in bore size and an 88mm/3.468inch stroke. Their 2381cc engines have 92.8mm/3.653inch bore size and a 88mm/3.468inch stroke.

Holbay of Grundisburgh also make a 2430cc capacity engine to order which has a 92.8mm/3.653in bore and a 90.0mm/3.544in stroke. They will also make to order a 2498cc capacity combination. This short block assembly is comprised of a 94.0mm/3.703inch bore and a 90.0mm/3.544inch stroke crankshaft. This size of bore cannot be accommodated by the standard cast iron block. The block used is the Diamond aluminium block which Holbay buy in from Millington Engineering.

Holbay make steel and aluminium flywheels to suit any crankshaft (nine bolt or six bolt retention and one to three dowels). Their lightweight aluminium flywheel has a steel insert which the clutch plate runs on which is held in place in the flywheel by 12 high tensile set screws. The ring gear is a shrink fit onto the flywheel, as well as having three cap screws holding it in place. When flywheels are made to suit Holbay's own crankshafts they are six bolt retention, with three dowel holes drilled in them to suit Holbay's three dowel hole system.

CLUTCHES

The standard 2000 engine's pressure plate and clutch plate are quite strong and will take a considerable amount of punishment. For competition use, though, the pressure plate and clutch plate must be changed for uprated components to avoid slipping clutch problems and driveline failure.

For the smaller engines, the 2000 engine's flywheel is a direct fit. The 1600 engine, for example, has a clutch

dowel/s need to be chamfered to ease fitting the flywheel.

Note that one well-fitting high tensile steel dowel is usually quite sufficient for racing applications. The dowel *must* have a 'press fit' into the crankshaft and a 'minimum clearance fit' into the flywheel (the flywheel will have to be tapped home).

Large capacity engines

Holbay make a range of crankshafts, forged connecting rods and forged piston combinations (machined from solid billet) to make larger than standard capacity engines. Their 2086cc engine has a 92.8mm/3.653in bore size and a standard 77.1mm/3.036in stroke plate and pressure plate (clutch cover) unit which is 7in in diameter, while the 2000's is 8.5in in diameter with a smaller inside diameter size (meaning a lot more surface area). The 2000 engine's flywheel can be lightened to provide a reasonably lightweight clutch, pressure plate and flywheel combination which is strong enough to take a lot of punishment but uses inexpensive standard parts.

For 2000 engines used in competition, uprated pressure plates are available that give considerably more clamping pressure. These pressure plates and clutch plates are available from the likes of Centreforce, Maxtorq, AP Racing and Sachs, to name but four, and will give excellent service. A single plate clutch is almost always sufficient. Consider using a 'Cerametallic' paddle clutch, however, as this type of clutch plate is almost indestructible, even when being severely abused.

If an uprated pressure plate is used in conjunction with the standard Ford clutch cable, the cable sheath simply collapses/concertinas and the clutch will not release. A genuine RS 2000 clutch cable will not collapse, but note that the outer sheath of the RS cable is spiral wound and needs a large curve when fitted.

When a high pressure competition clutch pressure plate is fitted, the pedal ratio needs to be a minimum of 5:1 for ease of operation – otherwise it will feel like pressing a brake pedal when changing gear! These high pressure plates are very strong and will not slip even under the most arduous applications.

If the application requires an even stronger clutch, a twin plate clutch assembly, such as an AP Racing one, will have to be fitted. Twin plate clutches are expensive and are rarely really necessary.

Chapter 3
Replacement parts

The choice of new replacement parts depends entirely on what the engine is going to be used for. Any road-going engine modified for high-performance will almost always be quite reliable with standard parts fitted into it (such as pistons, connecting rods, valves and rockers, for example). Competition engines are another matter, and it's definitely necessary to fit heavy duty parts capable of withstanding the rigours of the particular application.

PISTONS

The choice here is between standard cast pistons and forged pistons. The standard Ford cast pistons and the aftermarket cast pistons available (such as AE) are quite strong and are reliable up to 7000rpm on a continuous basis and 7500rpm on an occasional basis. Consider 7000rpm to be a safe engine speed for standard-type cast pistons fitted to a road-going engine.

Caution! For optimum results (i.e. maximum efficiency), the pistons must be brand new and fitted into on-size bores with the correct piston to bore clearance. Ideally they should be replaced at roughly 30,000 mile/50,000 kilometre intervals (to prevent piston failure through fatigue).

Cast pistons are quite suitable for any naturally-aspirated road-going engine because the pistons are seldom really stressed, the majority of their work being done at low to mid range rpm and not continuous revs at or above 7000.

Piston to bore clearance for a cast piston needs to be a minimum of 0.055mm/0.0022in for the smaller bore engines (1600 & 1800) and a minimum of 0.063mm/0.0025in for the larger engine (2000cc). Consider 0.075mm/0.003in to be the maximum piston to bore clearance for a cast piston.

Forged pistons

Forged pistons (Omega, Accralite, Holbay and Mahle) are reliable under extreme conditions, and most will withstand revs well in excess of 9000rpm.

Some forged pistons are lighter than others, but all forged pistons are strong. Consider forged pistons as being necessary for racing applications. Forged pistons require a greater piston to bore clearance and will need between 0.10-0.137mm/0.0040-0.0055in piston to bore clearance.

Caution! Note that some forged pistons require even greater piston to bore clearance (0.137-0.150mm/0.0055-0.006in) – always use the piston manufacturer's specified piston to bore clearance. Relevant specifications should be enclosed with the piston set or be printed on a label stuck on the box. The piston to bore clearance must be supplied with any new piston set along with the correct ring gap dimensions. Ensure that this important information is available at the time of purchase.

Note that large piston to bore clearances allow the piston more movement (rock), especially when the engine is cold. The rings are not held

SPEEDPRO SERIES

as square to the bore wall as they would be by a cast piston.

Piston pin retention can be by the standard method (an interference fit, which is the case for the standard connecting rod regardless of piston type) or, alternatively, the piston pin tunnel of the connecting rod can be honed out to allow the pin to be a floating fit and wire circlips (standard with forged pistons) used to retain the pin.

PISTON RING SETS

The type, size (meaning ring thickness) and quality of the standard Pinto compression rings is good and they are suitable for all applications for which the strength of the standard piston is adequate. In fact, the ring section is quite thick, but this is acceptable. The standard oil control ring is a three-piece component and is of good quality.

The minimum ring gaps for standard pistons are 0.38mm/0.015in for the two compression rings and 0.40mm/0.016in for the oil control ring. Maximum gaps are 0.50mm/0.020in for the two compression rings, and 0.625mm/0.025in for the oil control ring.

All forged pistons use different thickness compression rings so, once the move to forged pistons has been made, a standard ring set will not fit. Forged pistons always have premium grade rings fitted to them, which will be more expensive than standard rings, and may have to come directly from the piston manufacturer or an agent. Oil control rings are almost always three-piece sets, much the same as for the standard piston, a few comprise a one-piece ring with an expander. Ring gaps are as advised by the piston/ring manufacturer.

CRANKSHAFT BEARINGS

Heavy duty main and big end bearing shells are available for Pinto engines. For all applications using up to 6500rpm, the standard plain bearings are excellent. For competition applications using over 6500rpm, Clevite 77 lead indium bearing shell inserts, for example, should be used. In recent years there have been a few changes in the bearing shell industry, with the likes of Clacier/Vandervell being taken over by Michigan Bearings. Clevite 77 engine bearings are simply excellent. Repco in Australia ended up being called ACL after a management buy out.

Pinto engines are very strong and crankshaft bearing failures are uncommon. The diameter and width of the bearing journals ensures excellent reliability. Main bearing clearances should be 0.062mm/0.002in. Main bearing clearances should be 0.051mm/0.002inch to a maximum of 0.063mm/0.0025inch, and the connecting rod bearing clearances should be 0.048mm/0.019inch to 0.056mm/0.0022inch. Undersize main and big end bearing shell inserts are available from AE, ACL, Clacier/Vandervell and Clevite in minus 0.25mm/0.010inch, 0.05mm/0.02inch, 0.75mm/0.030inch and 1.00mm/0.040inch sizes. The crankshaft end float should be between 0.125-0.20mm/0.005-0.008inch with 0.005-0.006inch being ideal. **Caution!** Avoid using a crankshaft which has more than 0.20mm/0.008inch of end float when assembled, this is too much.

CAMSHAFT BEARINGS

Two types are available. The early ones are white metal, the later ones bronze. The bronze bearings are far more resistant to wear. The front camshaft bearing will virtually never wear out as it is much wider than the other two bearings whether it be white metal or bronze. The bronze centre bearing will last three times as long (at the very least) as the early centre bearing while a bronze rear camshaft bearing will last about ten times as long as an early rear bearing. Use of the later bronze bearings is *essential* and they are direct replacements.

CAMSHAFT SPRAY BAR

Caution! Always fit a new spray bar during a rebuild. The pipe can become quite clogged, especially if the engine has not had regular oil changes. Spray bars are inexpensive.

CAMSHAFT KIT

These kits are good value for money because everything required is included at a reasonable price. Standard replacement kits include a new camshaft, new set of rockers, spray bar, valve springs, pivot pillars and a container of special lubricant. High-performance camshaft companies also market camshaft kits. Valve spring pressure needs to be matched to the rpm that will be used. Many camshafts are run with far too much valve spring pressure for the rpm being used, and this will wear out the camshaft prematurely.

Some standard replacement kits feature dual valve springs that exert the maximum standard rated 'over the nose' pressure (up to 175 pounds). The inner valve spring is short and does not have much effect at the fitted height so, as a consequence, the seated valve spring pressure is about 55 pounds. 175 pounds is too much for a standard camshaft engine which will seldom see more than about 6300rpm. With

REPLACEMENT PARTS

this sort of spring pressure, expect to see major camshaft wear and, to a lesser extent, rocker wear after 50,000 miles/80,000km or, sometimes, much less.

GASKETS

Apart from the head gasket, the standard gasket set is adequate for all applications. The standard Ford (or replacement part manufacturers') cylinder head gaskets will hold 10:1 to 10.5:1 compression, although some replacements (non-Ford) may not last more than a year, or two, doing so. The average standard-type head gasket is 1.6mm/0.065in thick when compressed.

A readily available head gasket at a reasonable cost is the 'blue' Felpro item which is capable of holding 12:1 compression; it's 1.0mm/0.040in thick when compressed. The Sierra Cosworth cylinder head gasket, made by Reinz, is available (at quite high cost) but will not fail when fitted to a naturally aspirated engine; this is 1.3mm/0.052in thick when compressed.

Caution! The cylinder head's gasket surface must be perfectly flat, as must be the cylinder block deck surface. Any engine that is being built for a high-performance application must have both of these surfaces re-machined to prevent gasket failure. It's no use just having the cylinder head remachined. If the block surface is not machined and, at a later stage, is found to distorted, it's a lot more trouble to have it remachined when assembled, compared to when the engine is stripped. Pinto engine blocks are very stable and do not 'move' but to remachine the block and cylinder head gasket surface is always good practice.

SPECIAL BOLTS

Aftermarket heavy duty connecting rod bolts are available and their use is strongly recommended. The original Group 1 heavy duty rod bolt is readily available; it looks similar to the standard bolt but has a waisted shank.

Flywheel bolts must be new standard ones at the very least, but heavy duty flywheel bolts are available and are recommended.

The standard main cap bolts are more than adequate for most applications but main cap stud kits are available (ARP).

The standard cylinder head bolts are quite adequate but alternative stronger and better cylinder head retention is available (ARP head stud kit).

ARP make heavy duty connecting rod bolts, flywheel bolts, head stud kits and main cap stud kits for these engines (readily available from Kent Cams).

AUXILIARY SHAFT BEARING

Although not a high wear component this bearing should be replaced if there is any appreciable wear whatsoever. This bearing controls the auxiliary shaft (jackshaft) and, if wear becomes excessive, the oil pump/distributor drive gear mesh can be affected. The standard replacement part is quite suitable for all applications.

SEALS

There are four main seals. The rear of the crankshaft, the front of the crankshaft, the front of the camshaft and the auxiliary shaft. These seals should *always* be replaced during a rebuild, when the engine is stripped for maintenance or whenever there is a leak attributed to the seal concerned. The standard replacement seals as found in all gasket sets are suitable.

VALVES

The standard valves for 1600, 1800 & 2000cc Pinto engines are of similar size and quality (two piece valves with the heads and stems friction welded together). However, the valve head diameters and stem lengths do vary. Standard or standard replacement valves are suitable for use up to 7000rpm. Triple valve stem grooves are a feature of the standard system and are used to promote valve rotation. The standard valves' triple grooves tend to wear quite rapidly under racing conditions or anything over 7000rpm use on a continuous basis. The standard triple groove valves are quite suitable for any high-performance road application because the use of high rpm (over 7000rpm) on a continuous basis is unlikely.

Large diameter valves (Group 1 size) are also available with triple groove stems and use the standard keepers. Although these large valves offer improved valve area, the triple grooves of the valve stem are a limiting factor to reliability at sustained high rpm.

Some aftermarket parts suppliers (Burton Power and Vulcan Engineering, for example) have stayed with the triple groove keeper system with their lines of one piece stainless steel racing valves and do not experience the severe 'chopping out' of the grooves to anything like the extent of a standard Ford or a standard replacement valve. Burton Power and Vulcan Engineering have their valves made with a very tight, but still non-locking, fit between the valve stem and the keepers. Both companies use the standard Ford keepers and valve spring retainers, but, when assembled, instead of there being approximately 0.075mm/0.003in clearance between the grooves in the valve stem and the keeper locks, there will be approximately 0.025mm/0.001in.

SPEEDPRO SERIES

The grooves on these valves still burr with use, however, and the valves will, likely as not, require the burred edges of the keeper grooves to be smoothed with a honing stone before the valves can be removed from the cylinder head.

Once the fit of the keepers in the grooves machined into the valves becomes loose, replace the valves and keepers. With the frequent maintenance and frequent replacement philosophy firmly in mind, this system is perfectly satisfactory with few failures in service. In fact, failure will only occur if the valves have been left in the engine far too long.

Many racing engines are fitted with aftermarket one piece stainless steel triple groove racing valves which are made out of 214N material and are used to well over 8000rpm. The triple groove collet/keeper system is designed to promote valve rotation in a road going engine and is a system not normally associated with racing engines. Over time nothing can save this system from high wear at high rpm.

What can be done to improve reliability and reduce unnecessary wear is to reduce the clearance between the triple groove keepers and the valve stem grooves by removing a very small amount of material from the matching faces of each keeper. This is done by hand by rubbing each collet/keeper in turn back and forth over a sheet of 180 grit wet and dry paper placed on a flat surface. The removal collectively of 0.05mm/0.002in material from the matching faces of the keepers removes the normal clearance (valves are not able to rotate, or not easily). This procedure will tighten keeper fit to a degree, but the contact with the valve stem is not 360 degree, which would be better. Single groove valves and keepers are superior (more solid).

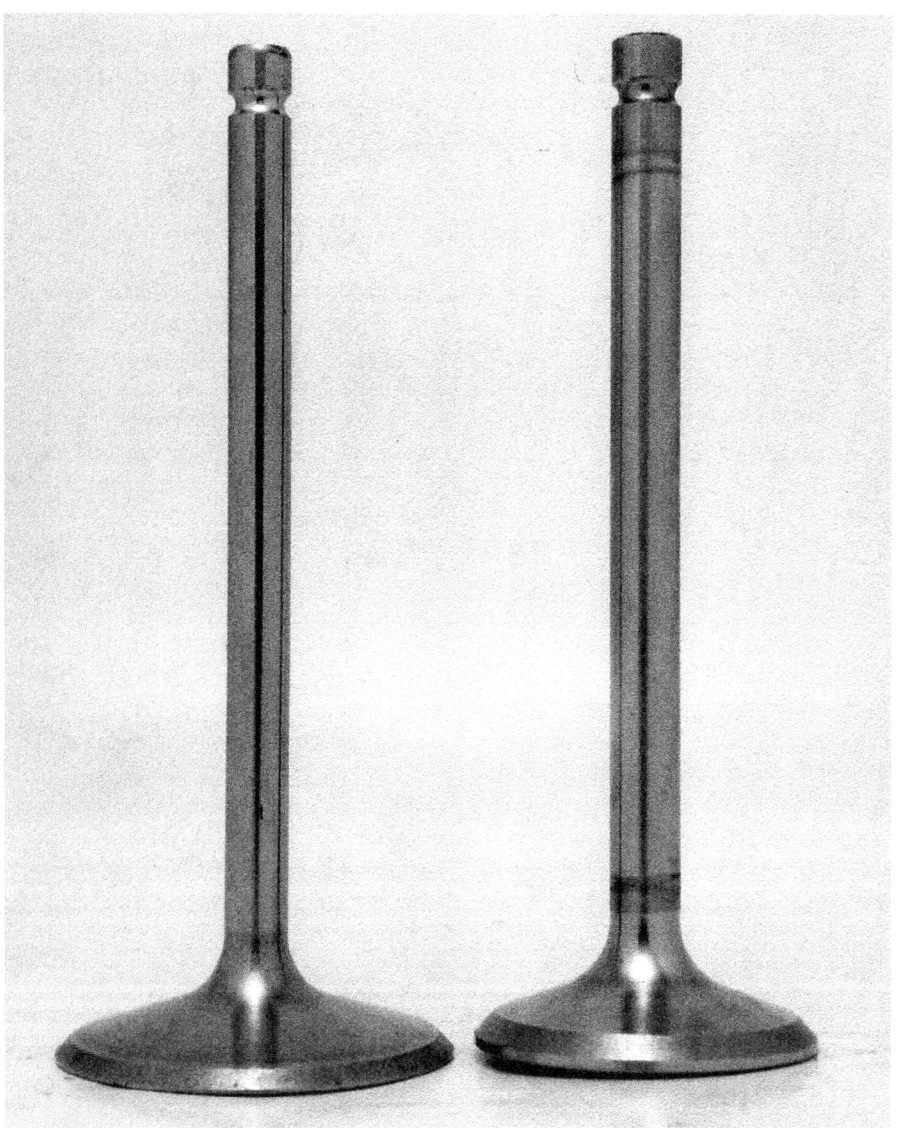

Holbay single groove exhaust valve on the left weighs 84 grams while the single groove inlet valve on the right weighs 70 grams.

Competition valves

Competition valves and keepers which have single grooves, do not rotate and can be regarded as being near unbreakable. Holbay only make single groove valves. These are usually installed in cylinder heads using replaceable valve guides (which Holbay make) to achieve the valve guide bore reduction necessary. This means machining the cylinder head to take these replaceable guides. The cylinder head's original valve guides can also be K-Lined down to suit 5/16inch, 9/32inch valve stem diameters. Holbay valves are all standard length (that's 2000cc engine length). Holbay make 1.750inch/44.46mm head diameter inlet valves with either 8mm diameter

REPLACEMENT PARTS

Components of a dry sump pump appropriate for Pinto/Cosworth.

stems or 9/32inch diameter valve stems and exahust valves with 1.500inch/38.1mm head diameters with 8mm diameter stems or 5/16inch diameter stems. Machined from solid bar keepers are high tensile steel, whilst the machined from solid bar retainers are available in titanium or high tensile steel.

VALVE RETAINERS AND KEEPERS

The standard valve retainers are solid and generally trouble free, but they are not light. Whenever standard valve spring retainers are being used in a high performance engine they should be brand new ones (old valve spring retainers have been known to crack on occasions!).They will withstand having the spring base remachined (1.2mm/0.048in maximum) to reduce valve spring tension, if this is required, and still be strong enough.

When an engine is being rebuilt, always fit brand new triple groove collets/keepers if triple groove valves are being used. They're not expensive and, although they seldom give trouble (crack), their replacement is accepted good practice.

VALVE STEM SEALS

Always fit valve stem seals to all valves to prevent oil contamination of the inlet charge and high oil consumption via leakage into the exhaust port. For all single valve spring applications, the standard valve stem seals will fit. These seals should also be replaced whenever the engine is stripped for maintenance. The standard early seals will not always fit when dual valve springs are used as the diameter of the standard seal is quite large (18mm/0.700in), later style standard seals usually will fit and there are plenty of other smaller outside diameter valve stem seals from other engines that will fit and give the required clearance (check with an engine machine shop for suitable seals).

OIL PUMP AND OIL PUMP DRIVE

A new oil pump and oil pump drive is *essential* in the interests of reliability and obtaining maximum oil delivery. The standard oil pump will supply oil at approximately 45psi to a correctly clearanced engine and is suitable for the majority of applications using up to 7500rpm.

The standard pump will not supply sufficient oil/pressure to an engine that has large crankshaft bearing clearances and is turning high rpm (over 7500rpm) continuously. The standard oil pump can have the relief valve spring packed by 3-4.7mm (1/8-3/16in) to increase the oil pressure, but this is a marginal solution: the fitting of an oil pump which has more capacity is the better option.

Capacity and pressure are not the same thing. A standard capacity oil pump can have its pressure increased by packing the relief valve up as much as possible and still not supply enough oil to the engine to maintain the desired amount of pressure. The pressure relief valve can limit the oil pressure, but only if the pump is supplying more oil than the engine can use.

High volume (capacity) oil pumps which have a taller bi-rotor (to increase capacity) and a relief valve set to a higher pressure are available but, for the majority of high-performance applications, the standard pump is adequate. When choosing a pump, consider 65psi (hot) to be sufficient maximum oil pressure for any wet sump Pinto engine (adequate for 8000rpm). Note that, as a rule of thumb, 8psi more oil pressure is needed for every extra 1000rpm of engine speed.

Always pack the new oil pump with petroleum jelly (Vaseline) before installing it in the engine. This way the pump is primed and oil will be drawn into the pump and fed to the engine almost the instant the engine is first turned over on the starter.

SUMP (OIL PAN)

The standard wet sump (oil pan) – preferably with a windage tray and baffles – is acceptable for most high-

SPEEDPRO SERIES

performance applications but *definitely* not for motorsport.

For motorsport applications a custom made sump *must* be made using a standard sump as a base. The sump's capacity needs to be increased over standard but, just as importantly, the oil reservoir must be correctly shaped (square, if possible, with dimensions of 230 by 230mm/9 by 9in and a depth of 80mm/3in) to ensure that the oil pickup is always immersed in oil. Engine bearings require pure oil, not aerated oil. The oil pickup must be centrally situated and 3mm/0.125in off the bottom.

If the sump base is quite close to the ground (less than 100mm/4in) the base should be reinforced and/or made of 2mm/0.80in thick panel steel. The reinforcement can be 25.4mm x 6.3mm/1 x 0.25in strapping welded front to back on the base panel of the sump; alternatively, the base panel can be corrugated or ridged to give it high strength.

The sump should also be fitted with a windage tray (set at the oil level of the reservoir) and a baffle which extends down into the oil from the underside of the windage tray. The windage tray must be clear of the connecting rod bolts at their lowest point by a minimum of 6mm/0.25in (70mm/2.75in down from the sump gasket rail for all Pintos). A windage tray acts as a barrier between the oil in the reservoir and the returning oil that drops down from the crankshaft bearings and pistons when the engine is running. Instead of the churned up returning oil aerating the oil in the sump reservoir, it first hits the windage tray, runs over the tray and back down into the oil reservoir in a controlled manner. The windage tray's baffle serves to slow the movement of oil in the reservoir when it's subjected to cornering, braking and acceleration G-forces.

Despite the wide variety of vehicles in which the Pinto engine is used, the custom made/modified sump should be made as close to the size previously given as is possible. The oil reservoir of the sump can be positioned at the front, centre or rear of the sump (depending on application), but its bottom should be parallel to the ground when the engine is in the vehicle. Note that most engines in cars are angled downwards front to rear; engines in racing cars are as parallel to the ground as possible.

The Ford RS 2000 came fitted with a cast aluminium oil pan which had a reasonable capacity, but good ones are getting hard to find and many have been cracked at some stage and repaired by welding. The Sierra Cosworth oil pan (alloy with good capacity) will also fit the Pinto block. Note that the RS sump will fit in Escorts but the Sierra Cosworth sump will not.

DRY SUMP

Only necessary for competition, this is the ultimate in oiling systems and can supply constant pressure up to 70-80psi. Fitting such a system involves the removal of the original oil pump and its replacement with toothed belt-driven external oil pump and twin scavenge pump pack.

The oil is fed into the main oil gallery in the normal manner but the galleries in the block which go from the oil pump to the primary side of the oil filter are bypassed. The cylinder block gallery drillings are just left and are not blocked off. The pressure pump gets its oil from a main oil tank and feeds it directly to the primary side of the oil filter via an adapter that is interposed between the block and the oil filter.

This adapter serves two functions: the first is to block off the original main oil gallery and the second is to feed oil from the pressure pump to the primary side of the oil filter. The oil goes through the filter and into the main oil gallery of the engine as normal.

A special sump with two scavenge pickup pipes (connected to the two scavenge pumps) is fitted. The oil that the scavenge pumps pick up from the dry sump pan is returned to the main oil tank which feeds the pressure pump before being circulated back to the engine. It's a relatively simple system which, all things being equal, will never see the pressure pump starved of oil. A dry sump system relies on the main oil tank being of the right capacity and of such a design that the oil it holds can be fed to the pressure pump under any circumstances including when high G-forces are in operation (braking, cornering and acceleration).

These dry sump oiling systems are not cheap but, if correctly matched to the engine requirements, offer 100 per cent oiling 100 per cent of the time. Burton Power and Holbay can supply dry sump lubrication systems for these engines.

TIMING BELT

Always fit a top quality new timing belt during the engine build and then, in high-performance applications, replace it more frequently than the engine manufacturer recommends. Belts do deteriorate and when they break the valves often get expensively bent, especially if the camshaft has more lift than standard. Using a camshaft with moderate extra lift, and pocketing the tops of the pistons to allow sufficient valve head clearance to preclude piston to valve contact, is a well-founded practice.

REPLACEMENT PARTS

Slightly wider than standard, higher strength (competition) timing belts are available from the likes of Burton Power (part number FT1045A). These belts have carbon-fibre in them and are much more durable than standard belts.

Burton Power can be contacted by phone on 0208 554 2281 or fax 0208 554 4828, or by email at sales@burtonpower.co.uk. They will ship parts to anywhere in the world and are an excellent firm to deal with.

CAMSHAFT AND AUXILIARY SHAFT THRUST PLATES

These two items wear out and allow the camshaft and auxiliary shaft more axial movement than standard. These thrust plates locate in grooves in the camshaft and auxiliary shaft and are not always a good fit even when new, let alone when worn, so, if any wear is present (measurable with a micrometer), replace them.

Chapter 4
Short block rebuild

A bare Pinto 2000cc cylinder block.

The fitting of top quality parts does not guarantee that an engine will be powerful and reliable. Parts *must* be fitted correctly: even the best parts, when fitted incorrectly, will fail and this is especially true for competition engines. *Nothing should be taken for granted and everything should be checked.* All parts *must* be check-fitted to prove that the 'running fit' or working relationship of all the parts is exactly as it should be. If you want maximum power and reliability, there can be nothing hit or miss about engine assembly – *do it right!*

There are no bad Pinto cylinder blocks, but there are cylinder blocks which can be regarded as better than others. The later Ford Sierra 2.0IS block (which has a small '205' marking) has thicker sump rails than the earlier blocks (Cortina, etc) and much the same as the Cosworth 205 block (which has a very large '205' marking). There are no serious cylinder block strength problems and, basically, any 1970 on block can be used. Sierra blocks (1986 on) have '165' cast in them (in large numbers) for 1600cc engines, '185' (in large numbers) for 1800cc engines and '205' (in large numbers) for 2000cc. The bore wall thickness is the same on all Pinto blocks.

PERMISSIBLE BORE OVERSIZES

Pistons to a maximum oversize of 0.060in are available for the 1600 and 1800 engines, but the recommendation is to bore any block to the *smallest* oversize that can be used. Except in extreme cases, there is little point in boring a block to the maximum size within the standard range. Boring a block oversize *does not* result in any significant increase in the power that can be developed by the engine: real power comes from

SHORT BLOCK REBUILD

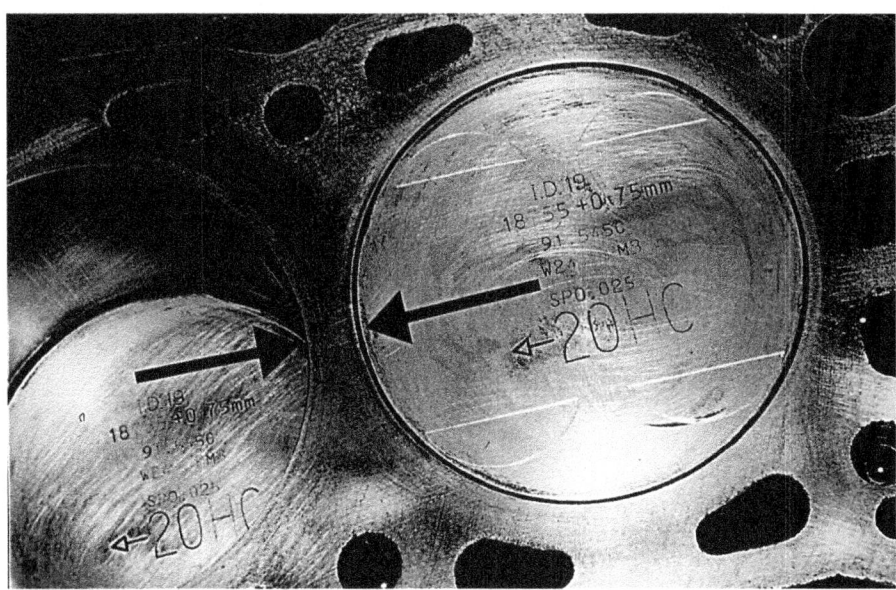

Pinto block deck. Arrows show the area where the bore walls are thinnest because of the waterways between the bores.

hold the cylinder head correctly and without distortion. In view of this a 'stock block' without sleeves and with a minimum overbore is always best because the bore walls are stronger and not prone to flex (poor piston ring sealing) or undue distortion under pressure causing head gasket failure.

Basically, all Pinto blocks, from first to last, including the Sierra Cosworth or later Sierra IS, have bore walls that are approximately 6mm/0.240in thick when standard, except in the area between the bores where the bore wall thickness reduces to approximately 4.0mm/0.160in.

modifying the engine correctly and using the right camshafts, carburation, porting, compression ratio, and so on.

For the 2000cc engine, pistons of 2.25mm/0.084in oversize are available. These pistons give a capacity of 2094cc and this is the largest capacity you should ever consider because, at this size, the bores are still just within acceptable limits (bore wall thickness) and all of the commercially available cylinder head gaskets will still fit without trouble.

REPAIRED BLOCKS

Blocks at maximum overbore or with cylinder damage are frequently repaired by sleeving the bores, but for high-performance use there is always some risk attached to this method for two reasons. Firstly, if the sleeve is too thin in wall thickness (1.25-1.5mm/0.050-0.060in) there is considerable risk of lengthwise cracking because sleeves really need to be a minimum of 2.0mm/0.080in thick to avoid this happening.

Secondly, if the original bore is bored out too much in order to accommodate a thick-walled sleeve, there is minimal material left in the block to hold the sleeve firmly.

The bore walls of cast iron blocks are structural in that they work in conjunction with the engine's outside walls and the deck of the block to

COMPONENT INSPECTION

With everything stripped and all components thoroughly cleaned, inspection can begin. *Only components that have been cleaned to bare metal can be inspected adequately.*

BLOCK & MAIN BEARINGS

With the block and main caps cleaned thoroughly using thinner to remove all traces of oil and residue, visually check

Main cap sitting on the block register ready to be tapped home.

SPEEDPRO SERIES

Cap being tapped down into the register of the block.

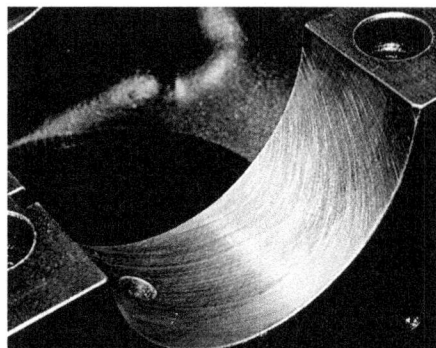

Main bearing tunnel bore surface as machined by the FoMoCo. Nothing less than this finish will do.

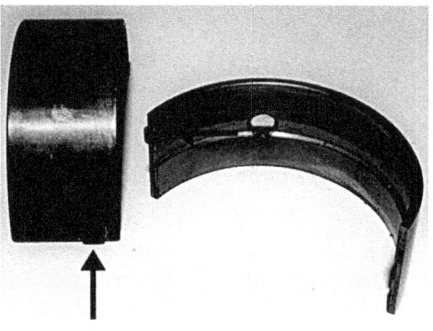

Location tabs on main bearing shells (inserts).

all surfaces for cracks. This includes 100 per cent of each bore's surface, the main caps, main cap webs, the block's deck surface and the area around all tapped holes. Cracks are frequently quite easy to see. Engines that have only ever been used in road-going vehicles seldom have cracked blocks. If there is an obvious major crack, the block is a write-off and further checking is pointless.

Main bearing caps

The main caps fit into the block rather than onto the block. This fit is by way of a machined register in the block in which the cap is a tight fit. If, instead of needing to be tapped in, a cap simply falls into its register, the block is *not* suitable for further use.

With the register and the base of the main cap scrupulously clean, the main cap is positioned onto the block (with the arrow pointing to the front of the block) with one edge of the cap located in the register while the other edge of the cap is up on the other register.

The cap is held with a definite bias toward the left and the top right-hand side of the cap tapped downwards so that the cap 'snaps' into the block's register. Two or three very light taps with a small copper hammer or rawhide hammer is all that it will take. With the cap correctly located the two main cap bolts are oiled, then screwed in and torqued to the correct tension. Note: it is

Main bearing tunnel bore being measured in the vertical plane and at 70 degrees to the left and right of the vertical plane.

SHORT BLOCK REBUILD

assumed that the threaded holes in the block were thoroughly cleaned out and that the threads of the bolts were thoroughly cleaned, too.

With the cap fully torqued, the tunnel aperture is measured with an inside micrometer and the size checked against the manufacturer's specifications. Ford list the main bearing tunnel bore diameter tolerance (range of acceptable size) as 60.620-60.640mm/2.390-2.395in. The tunnels *must* be within this range. The small measurement is known as 'bottom size' and the larger as 'top size.' The optimum size is bottom size. Reject any main bearing tunnel bore measurement over top size, but note that blocks can be remachined by an engine reconditioner to restore optimum, or acceptable, main bearing tunnel bore sizes.

The first of three measurements is taken in the vertical plane and in the middle of the main bearing tunnel. The second measurement is taken 70 degrees left of the vertical plane and the third measurement is taken 70 degrees to the right of the vertical plane. With three measurements taken in these places the integrity of the main bearing tunnel can be clearly ascertained.

The surface finish of the main bearing tunnel bores should be very smooth (the same as a freshly honed cylinder bore with a cross hatch pattern). There should be no marks indicating a spun (bearing shells have rotated in their housing) bearing and, if there are, even if the tunnel measures as being in tolerance, the block will have to be align bored or align honed to remedy the situation.

Check-fitting main bearing shells

With the integrity of the block's bearing

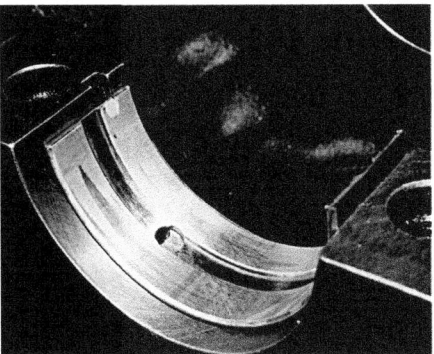
First step in fitting a bearing shell. The location tab is positioned in the machined groove 3mm/0.25in down from the mating surface.

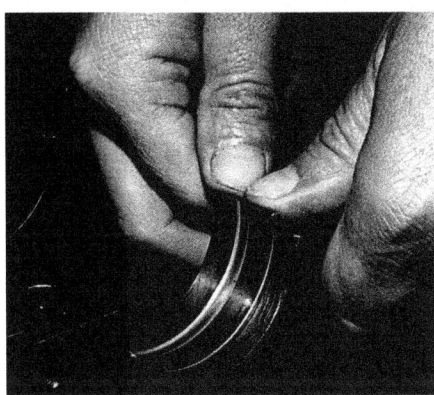
How the bearing shell should be pushed down into the tunnel.

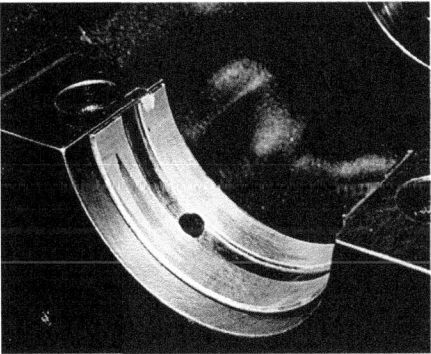
Shell correctly positioned in the block with the ends at equal heights.

tunnels confirmed as being 'on size' (at bottom size or top size or in between) the new bearing shell inserts that are

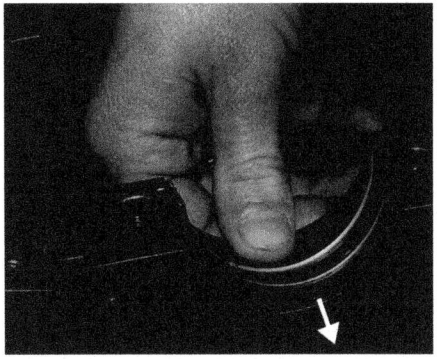
The easiest way to remove a shell is to push it with your thumb in the direction arrowed.

going to be used in the engine at final build *must* be check-fitted to the block. Bearing shells are very accurately made, but, as with all mass produced components, there is always some dimensional variation (however small).

In the first instance the bearing shell inserts are randomly paired up and fitted to a main bearing housing. Each shell/cap combination is then checked for 'bearing crush' to see just how tightly the bearing shell inserts are actually held in the main bearing tunnel. The bearing shell inserts can be mixed and matched to a certain degree to get an even and acceptable amount of bearing crush.

The usual procedure is to start at the rear of the block and work forward. This usually proves to be most convenient, especially if the engine block is being held on an engine stand, as the main bearing furthest away from you is worked on first.

Once a pair of bearings shells have been check-fitted they *must* be labelled as belonging to the particular main bearing tunnel in which they were checked.

Clean the new bearing shell thoroughly using thinner and soft paper towels to remove all traces of dirt or protective grease from the backs of the inserts and the bearing surfaces. The bearing tunnel halves are also cleaned (block and cap).

The grooved bearing inserts are fitted into the block and the plain bearing inserts are fitted into the caps. The bearing shell inserts are fitted into the block or cap with the location tab of the bearing shell insert positioned in the machined groove of the block or cap. Note that the tab of a bearing shell insert is for location purposes only and is not a device designed to stop bearing shells spinning in their tunnel. If the main bearing tunnel size is too large or the bearing shell inserts are too small, or a combination of both, nothing is likely to stop a main bearing 'spinning up' when the engine is subjected to stress.

The bearing shell insert is fitted into the main bearing tunnel by first locating the tab into the machined groove in the block or cap approximately 3mm/0.125in down from the mating surface of the block/cap. With the bearing shell insert accurately located, the end of the shell that is above the mating surface of the block/cap opposite the tab is pressed downwards with fingers until it is nearly flush. The ends of the bearing shell insert will usually be slightly proud of the mating surfaces of the block and cap, but not always. The reason for this is that the actual positioning of the partline of the block and caps is not necessarily exactly on the true centreline of the crankshaft's axis. Position the ends of the bearing shell inserts equally in relation to the mating surface of the block or cap.

With the bearing shell inserts correctly positioned in both the block and the cap, the cap is then placed into the block register. The cap will be sitting up on one edge of the register and must be snapped into the register as previously described. The cap bolts

Shell location tab positioned below the mating surface of the cap just prior to being pressed into the cap.

Shell correctly positioned in a cap with the ends at equal heights.

are then fitted and torqued to the correct value.

An inside micrometer is now used to measure the bearing bore tunnel size to check that the bore diameter is in tolerance. Care *must* be taken when using an inside micrometer on soft material such as bearing surfaces so as not to get a false reading. The anvils of the inside micrometer should lightly rub over the surface of the bearing. **Caution!** They *must not* dig in to the material and mark the bearing surface.

The bearing bore may not be exactly circular so there can be some slight variation (larger) from the vertical measurement in the measurements taken at the 70 degree measurement points. Go by the vertical measurement.

Big end bearing bore being measured in the vertical plane and at 70 degrees to the left and right of vertical.

With a bearing shell tunnel bore measured in three places, the overall shape and size of the tunnel is ascertained. When the required bearing clearance is deducted from the measurements, the resulting dimension should be the same as the intended crankshaft journal size; if it's not, the crankshaft journals will have to

SHORT BLOCK REBUILD

be eased. If the crankshaft journals are all to be reground then each individual journal should be ground to suit the size of the bearing in which it will run (within factory tolerances).

With the fitted sizes of the main bearing bores measured and checked against the journal sizes, the precise running clearance is known (record all sizes for future reference as, when the engine is stripped for its next rebuild, the amount of wear in relation to usage can be accurately ascertained). The main bearing clearances should all be within the factory tolerance of 0.5-0.064mm/0.002-0.0025in. Note that for high-performance engines the optimum clearance is at the higher end of the tolerance; if required the desired clearances can be obtained by individual machining of the crank journals.

Most engines will be found to have bearing clearances within the factory tolerances. Clearance problems can sometimes be remedied by another set of bearing shells, or the affected crank journals can be machined, within permitted tolerance, to the appropriate sizes.

Checking bearing crush

With the bearing shells fitted and the two main cap bolts torqued, the bearing shell inserts are forced to conform to the tunnel size of the block. By checking the 'bearing crush' the quality of bearing fit is ascertained. Undo and remove one bolt (either one) and, starting with a 0.075mm/0.003in feeler gauge, see if the feeler gauge will go into the gap between the cap and block.

By using a range of feeler gauges the exact size of the gap can be determined. The allowable range is 0.075mm/0.003in to 0.015mm/0.006in.

Main cap with one bolt removed. The other bolt is fully tensioned. Measure the bearing crush gap with feeler gauges.

If no gap appears when the bolt is released there is something wrong with the bearing shells, they will, effectively, be loose because there will be no bearing crush present. Such a situation is *not* acceptable – try another set of shells.

If the gap is over 0.20mm/0.008in it means there is too much crush, which is unacceptable because there is likely to be a resulting distortion at the partline. The set of shells will have to be exchanged for another. Alternatively, by mixing and matching the existing shells the desired bearing crush might be achieved. If there is a variation between the bearing shell combinations in each main cap, take one shell from a main cap that has a large crush height (0.15mm/0.006in plus) and change it with a shell from a bearing shell combination from another main cap that has a small crush height.

The ideal situation is to have all main bearing shell combinations with 'crush heights' of 0.10mm/0.004in to 0.125mm/0.005in.

If there is insufficient bearing crush present the result is usually a spun bearing and, if this happens, serious and expensive engine damage will result.

CONNECTING ROD CHECKS

All connecting rods *must* be checked very thoroughly. The checks involve a thorough crack test, a straightness check in both planes, a length check, hardness testing, piston pin bore diameter size check, connecting rod bearing tunnel diameter size check and an out of round check. The old pistons are pressed off. This is done on a garage press *without* the use of any heat. The interference fit of the pin is usually 0.010-0.012mm/0.0004-0.0005in and this is well within the capability of any 5 ton press. Once sufficient pressure has been placed on the piston pin to create initial movement, the piston pin will continue to move.

Connecting rod crack testing

The connecting rods are crack tested by an engine machine shop. The connecting rod is magnetized, using an electromagnet, and a mixture of 'Magnaflux' powder and kerosene is sprayed over it. If there is a crack, the powder will congregate along the crack and 'black light' will highlight it. The operator must be skilled in the use of this equipment, as a crack can be missed by the inexperienced. Discard any cracked rods.

Connecting rod straightness

Having passed the crack test, each connecting rod is then placed on a connecting rod alignment jig which will check the alignment of the rod's crankshaft bearing and the piston pin bores in two planes to show whether the rod is bent or twisted. These jigs are very accurate and can check connecting rod alignment in both planes to within 0.003-0.012mm/0.0001-0.0005in.

SPEEDPRO SERIES

Connecting rod alignment jig. Here a rod is having alignment in both planes checked at the same time. These jigs can pick up misalignment down to 0.0001in.

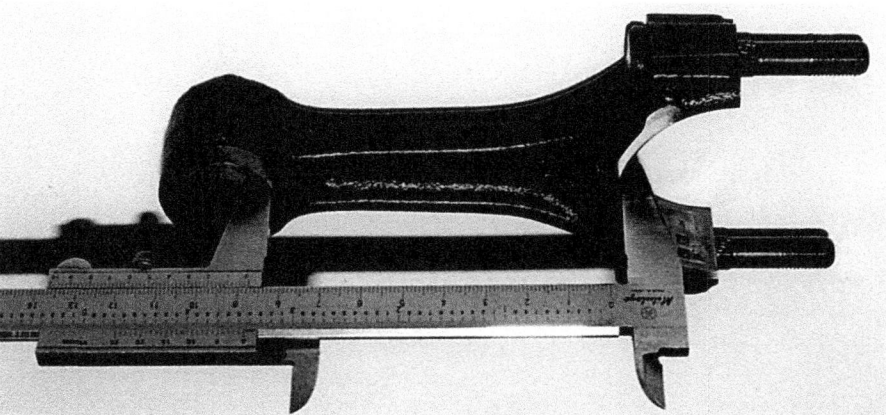

Connecting rod being measured for length.

While it is possible to straighten rods, for high-performance use it's not recommended. Throw distorted rods into the rubbish bin.

Connecting rod length

The alignment jig does not check the length of the connecting rod. Length measurement is made off the machine with a vernier caliper. The dimension is measured from the top of the connecting rod bearing bore to the bottom edge of the piston pin bore. The average measurement will be in the vicinity of 87.5mm/3.448in. If one rod is well down in size compared to the others, it's probably bent.

Actual connecting rod length is the distance between the centre of the connecting rod bearing tunnel and the centre of the piston pin tunnel. This distance can be calculated by adding together half of the diameter of the connecting rod crankshaft bearing tunnel (27.5mm/1.083in) and half of the diameter of the piston pin tunnel (12.0mm/0.472in) and then adding that to the distance as measured from the top of the connecting rod bearing tunnel to the bottom of the piston pin tunnel.

Rods are usually within 0.104mm/0.004in of each other but, if bigger discrepancies are discovered, piston crowns can be machined to even things up.

Connecting rod hardness

The piston pin ends of connecting rods can be damaged by the use of heat during piston pin removal. If you suspect a problem of this type, the connecting rods can be checked for hardness in an appropriately equipped machine shop. Hardness in the 'I' beam area is first measured and then the piston pin end of the rod in order to make a direct comparison. An overheated piston pin end will be well down on hardness (usually measured with a Rockwell hardness tester using the C scale). This test settles any doubt as to the integrity of the material strength of the connecting rod.

PISTON PIN TO CONNECTING ROD FIT

The piston pin end of the standard connecting rod is designed to have what is called a press fit/interference fit between the piston pin and the piston pin bore. This means that the piston pin has to be 0.012-0.020mm/0.0005-0008in larger than the piston pin tunnel in the connecting rod. The maximum amount of difference allowable between the piston pin and the piston pin tunnel of the connecting rod is 0.040mm/0.0015in. The majority of piston pin tunnel to piston pin fits will be 0.012-0.015mm/0.0005-0.0006in.

Measure the internal diameter of the piston pin tunnel in the connecting rod. If this fit is to be 100 per cent reliable in a high-performance application there must be an *absolute* minimum of 0.0127mm/0.0005in interference between the pin and the connecting rod. Ford list the standard amount of interference fit between piston pin and connecting rod as 0.018mm/0.0007in; consider more than 0.039mm/0.0015in to mean that the rod is a write-off. The nominal piston pin tunnel diameter of a standard connecting rod is 23.982mm/0.944in.

Note that before the new pistons are purchased, the connecting rods must already have been measured and checked so that when the piston set is

SHORT BLOCK REBUILD

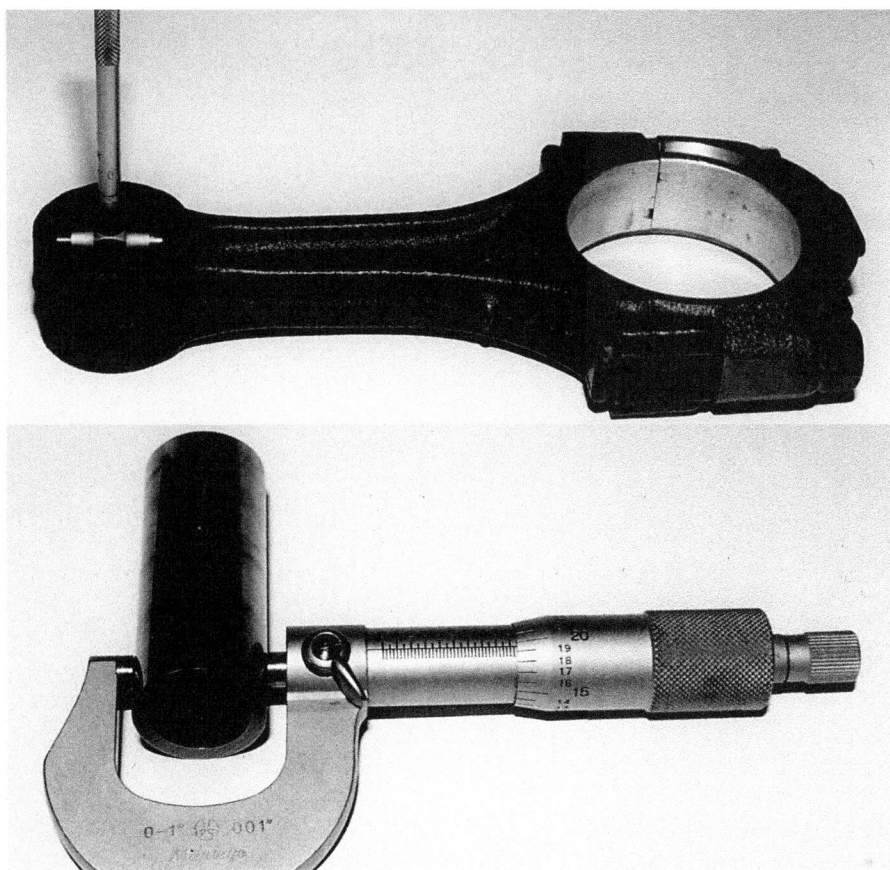

Small end tunnel of connecting rod being measured with a telescopic gauge. An outside micrometer is then put over the telescopic gauge and a measurement taken. Piston pin, too, needs to be measured.

piston pin tunnel diameters *before* the piston set is purchased: this way, the minimum size of piston pin acceptable will be known before the piston set is paid for and taken away. Piston pin sizes do vary by as much as 0.010mm/0.0004in so the amount of variation possible in 'interference fit' can be very significant.

If the inside diameter of the piston pin bore is no longer 'on size' the connecting rod will have to be replaced. The piston pin bore can be honed out oversize, but an oversized piston pin will have to be procured and the piston's piston pin bore will have to be honed out to suit the new pin. Generally, a new connecting rod is found in an effort to keep everything stock.

CHECKING CONNECTING ROD BIG END TUNNEL SIZE

Measure the rod's big end tunnel diameter using an inside micrometer. Measure in three positions – vertically (in-line with the I-beam of the connecting rod) and then at 10mm/0.375in above the partline of the connecting rod and the cap respectively.

The connecting rod big end tunnel diameter tolerance is listed by Ford as being 55.0mm/2.1653in (bottom size) to 55.02mm/2.1660in (top size). Consider the bottom size to be best for high-performance applications but, provided there is still sufficient bearing crush, having a top size tunnel diameter will not cause any problems.

If the big end tunnel is found to be misshapen, or just too large in diameter, the connecting rod will have to be resized on a connecting rod honing machine. In fact, it's good engineering practice to resize connecting rods when an engine is

bought, the minimum size the piston pins must have will be known. Check the sizes of the piston pins before paying for the pistons and taking them away. Piston pin diameters do vary so if there is not enough interference fit, yet the connecting rod is 'on size,' try another set of pistons.

The sizes of connecting rod piston pin tunnels can vary within the range of 0.012-0.040mm/0.0005-0.0015in which is a tolerance of 0.240mm/0.001in. The size of piston pins also varies (by as much as 0.010mm/0.0004in) so it is possible to end up with a piston pin and connecting rod piston pin bore combination that has virtually no interference fit. Check all piston pin sizes and all connecting rod piston pin tunnel diameters.

Match the piston pin sizes of the piston set to connecting rod piston pin bore diameters that give the maximum interference fit for each combination. By selectively matching the piston pins to the connecting rod piston pin bores, the maximum available interference fit will be present in each connecting rod and piston pin combination.

Caution! Note that piston pins are selectively fitted to each piston and they must *not* be swapped around. Measure the connecting rods'

SPEEDPRO SERIES

Connecting rod bearing tunnel being measured in the vertical plane and at 70 degrees to the left and right of vertical plane.

being reconditioned and especially so for high-performance applications. If used standard connecting rods are being refurbished for use in any high-performance engine, the recommendation is to resize them to bottom size as a matter of course. Before this is done the connecting rod bolts are replaced by new ones. This ensures the cap is located perfectly before the connecting rod is resized: if the connecting rod is resized and the bolts replaced later, there is no absolute guarantee that the cap is in perfect alignment.

CONNECTING ROD CRANKSHAFT BEARING TUNNEL RESIZING

High-performance engines must have the connecting rod crankshaft bearing tunnels (big ends) resized to ensure that the connecting rod bearing tunnel is perfectly round. Equally important is that the size of the connecting rod bearing bore diameter is on 'nominal size' or 'bottom size' to ensure that the two connecting rod bearing inserts are held in the connecting rod bearing bore with the maximum 'bearing crush' possible within the factory specifications.

A connecting rod bearing tunnel on 'top size' could see the bearing crush, as measured with a feeler gauge, down to zero or 0.024mm/0.001in but this is insufficient. The minimum amount of crush required is 0.104-0.150mm/0.004-0.006in.

The factory specifications for your engine will list a 'nominal size' with a plus and minus tolerance for the connecting rod bearing tunnel. The plus tolerance size is called 'top size' and the minus tolerance size is called 'bottom size.'

Driving out connecting rod bolt. Note how bearing cradle is clamped between soft jaw protectors.

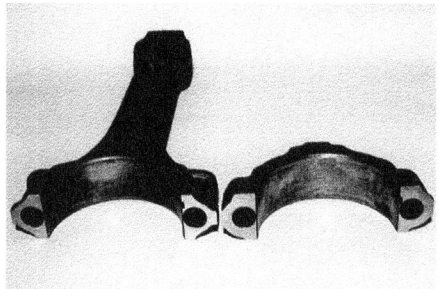

Cap and rod mating faces which have had material removed.

Removing connecting rod bolts

When connecting rods are resized the rod bolts are removed from the connecting rod and both the cap and connecting rod matching faces are reground. Removing the bolts can be quite difficult but, as the bolts have to be removed for replacement anyway, this is the time to do it.

The connecting rod bearing end of the connecting rod is held in a large vice which has protective shields fitted (tin plate or panel steel). With the connecting rod held as suggested, the bolts are driven out using a mild steel drift punch (25mm/1in diameter) which is hit with a hammer. The reason for using the drift punch is to avoid any possibility of hammer contact with the mating surface of the connecting rod as this could render the connecting rod useless.

SHORT BLOCK REBUILD

If the bolts are very tight, the connecting rod may start to move in the vice. Persevere with this method as it almost invariably works.

If you can't shift a bolt without danger of damage to the rod, take the offending rod/s to an engine machine shop; they'll drill out the centres and remove the bolts for you.

Refacing connecting rod and cap joint

During the regrinding of each connecting rod's mating face a minimal amount of material is removed, but always sufficient to clean up the surface completely. The amount of metal removed also depends on how much error there is in the connecting rod bearing bore: the amount removed to correct all problems is usually 0.024-0.050mm/0.001-0.002in.

Note that a connecting rod's bearing bore can be restored by removing material from the cap alone (without removing the connecting rod bolts) but, unless the matching surface of the connecting rod is machined, the integrity of that surface remains an unknown quantity.

Once the tunnel bore diameters have been reground, the edges of the connecting rod and cap mating surfaces are hand-filed to clean them up. Use fine needle files to do this. The edges should have a 0.12mm/0.005in chamfer when finished. The reason for doing this is that any burring of edges, or slivers of steel, can prevent the cap and connecting rod mating surfaces from matching properly and bearing 'crush' will be reduced – which could result in engine failure.

Fitting new connecting rod bolts

The new connecting rod bolts are fitted to each rod using an aluminium or copper drift punch and a hammer. For ease of fitting, the connecting rod can be placed on a piece of 25 x 25mm/1in x 1in aluminium alloy which has had two clearance holes for the connecting rod bolts drilled in it. The connecting rod sits on top of the aluminium square bar and as the holes in the bar are through holes, the bolts move through these holes as they are driven into the connecting rod. Aluminium is used for the assembly jig as it will not mark the very important mating surface of the connecting rod.

Make sure that the head of each bolt is positioned correctly. There must be *no* doubt that the head of the bolt is in contact with the surface of the connecting rod bolt head machined recess. The head of the bolt *must* be in contact with the connecting rod's broached (section of the shoulder cut away to give a flat seat for the bolt head) surface.

Fit the cap to the connecting rod and torque the nuts to the specified tension. It is acceptable to tap the cap onto the threaded portions of the connecting rod bolts (which must be parallel) with a copper hammer or drift. **Caution:** some sockets are not 'slimline' and will not clear the cap sufficiently and slight misalignment can result. Be *certain* that the socket you use fits over the nut but does not interfere with the cap!

Aluminium jig used during the fitting of connecting bolts. The jig is T-shaped in cross-section so that it cannot slide down between the jaws.

Correctly positioned bolt head left; incorrect, right.

Edges of connecting rod mating faces being chamfered with a needle file and the finished job. Chamfer, arrowed, should be 0.18mm/0.007in.

SPEEDPRO SERIES

Connecting rod bearing tunnel being honed.

Connecting rod bearing tunnel honing

The connecting rod bearing tunnel can now be honed to size at an engine machine shop on a specialized machine specifically designed for this operation. This process will see the connecting rod bearing bore rehoned to 'nominal size' or 'bottom size' and the resulting tunnel will be perfectly round without any distortion (ovality, humps or hollows).

AFTERMARKET CONNECTING RODS

The manufacturers of these connecting rods guarantee the accuracy of the product and new connecting rods do not normally need to be checked as the factory quality control system is virtually fail-safe. Even so, it is recommended that all components are checked for straightness, piston pin tunnel diameter and connecting rod bearing tunnel diameter. While expensive, these rods are very durable but, ultimately, all engine components have a 'fatigue life' and even the very best aftermarket connecting rods have a limit!

Used aftermarket connecting rods should be checked and rebuilt if necessary in just the same way as used stock-type connecting rods. This would include crack testing, straightness testing, connecting rod bearing bore resizing and the fitting of new connecting rod bolts.

All connecting rods (new or used) have to have the bearings check fitted into them to ensure 'bearing crush' is correct and that the inside diameter of the bearing tunnel is compatible with the crankshaft journal diameter. All other aspects of 'check fitting' these connecting rods into an engine block are the same as for the stock type of connecting rod. The piston pin retention method may vary by way of being a press fit or fully floating and the side clearance be slightly more than stock, but the basic principles of 'check fitting' remain the same.

CONNECTING ROD BOLTS

The old connecting rod bolts *must* be replaced by new standard items or, preferably, aftermarket high-strength bolts as part of the rebuild.

Caution! Any engine that is going to be revved above 6500rpm on a frequent basis *must* have aftermarket high strength connecting rod bolts fitted. Group 1 bolts, for example, are definitely stronger than the standard components and are rated at around 190,000 pounds per square inch tensile strength.

Caution! When new connecting rod bolts are fitted the connecting rod bearing bore will usually need to be resized. The reason for this is that when the old bolts are removed and the new bolts installed, the cap

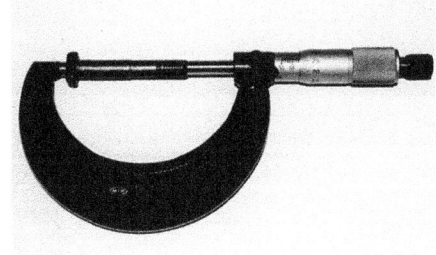

Measuring length of rod bolt.

Rod must be properly held in vice fitted with jaw protectors before the nuts are tightened to full torque. Move rod to get second nut in same position before tightening it too. Socket must not bind on the cap.

may not fit back in exactly the same position as it previously did. Obviously the new position of the cap will be very close to the original and the amount of error will be very small (0.003-0.024mm/0.0001 to 0.001in). However, it only takes one unchecked connecting rod bearing tunnel to be out to lead to engine failure ...

If a cap does not line up, the bearings will not be held in a perfect circle, or the bearing inserts may end up being held in the connecting rod bearing bore in slightly staggered positions. Either way, the situation *must* be rectified.

All connecting rod bearing tunnels should be remachined (honed) after the fitting of new bolts to remove all possibility of cap misalignment, tunnel bore distortion and to ensure the correct tunnel inside diameter size so that the bearing shell inserts are perfectly held.

Checking connecting rod bolts for stretch

Tightening nuts to a specified torque is just a convenient way of setting the bolt with a prescribed amount of

SHORT BLOCK REBUILD

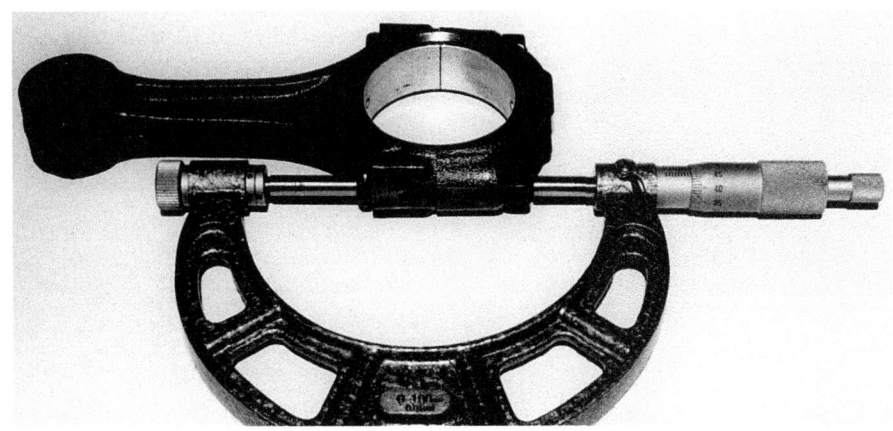

Measuring length of fully torqued rod bolts.

tension that correctly stretches the bolt within its elastic limits. This means that if the tension is removed from the bolt (nut removed) the bolt will return to its original size. Note that most bolt threads must be lubed before the nut is put on.

Caution: when tightening connecting rod nuts you *must* always check to see that the socket does not interfere with the side of the connecting rod cap. Check with each connecting rod and cap combination. Once satisfied that there is no contact, keep the socket used specifically for this operation so that there is no chance that a similar socket (which may not be clear of the cap) is used. Take *nothing* for granted and *double check* everything.

Measure all new bolt lengths as they are taken out of the packet and compare these individual measurements against the new lengths after torqueing the bolts. All bolts must take the prescribed torque and not stretch more than the specified amount. Expect the bolt to stretch 0.104-0.200mm/0.004in-0.008in (depends on the bolt). The correct amount of stretch for a particular bolt will be listed in the fitting instructions that come with the bolts.

Most aftermarket bolt sets come with torque settings which must be checked against the increase in the length of the connecting rod bolt. If a bolt takes the prescribed torque but stretches well beyond the listed limit, that bolt *must not* be used.

On odd occasions a new bolt will not take the torque setting because it stretches above a particular torque. For instance, if the torque requirement is 35 foot pounds, yet the bolt won't go over about 18 or 22 foot pounds, check the length of the bolt against its original size. Expect the bolt to be about 0.75mm/0.030in longer before you notice that it is not taking the torque. The most likely reason for this type of bolt failure is incorrect heat treatment. Replace the bolt.

CONNECTING ROD BEARING CRUSH

With rod tunnels resized to 'nominal size' or 'bottom size' and new connecting rod bolts fitted, the new bearing inserts can be 'check fitted' into the connecting rods and the amount of 'bearing crush' checked.

Thoroughly clean the backs of the bearing inserts and the tunnel surfaces of the connecting rods and caps using a solvent. Install the inserts in the cap and connecting rod, fit the cap to the rod and torque the nuts. Hold the connecting rod in a vice equipped with soft jaws (to protect the side faces of the connecting rod) and make sure the connecting rod is held as fully across the partline as possible.

The next step is to undo one nut to remove tension from that side of the connecting rod. The inserts exert a radial force which causes this side of the cap to lift. The gap is measurable with a feeler gauge and is termed the 'crush height'. Expect the gap to be anything from 0.12 to 0.15mm/0.005in to 0.006in. If the tunnel size is on 'bottom size' and there is 0.024mm/0.001in crush height the inserts are faulty (very rare, but possible).

If some connecting rod bearing insert combinations give varying crush heights, such as some having 0.015mm/0.006in and some 0.075mm/0.003in, mix and match the insert halves to see if the crush heights can be evened up a bit. There is frequently some slight variation between the inserts. This means swapping an insert half from

Cap 'sprung' by bearing crush. Feeler gauges inserted between the cap and rod should reveal a 0.1-0.15mm/0.004-0.006in gap.

SPEEDPRO SERIES

Connecting rod bearing internal diameter being measured in the vertical plane and at 70 degrees to the right and left of vertical.

a 0.015mm/0.006in crush height connecting rod into a 0.075mm/0.003in 'crush height' connecting rod combination which will alter the crush height, especially if the connecting rod tunnel sizes are all identical.

Checking connecting rod bearing internal diameters

With the crush heights checked and all connecting rods having similar feeler gauge measurements, the nuts are torqued again and the bearing bore diameters accurately measured with an inside micrometer. The inside diameter of the bearings is measured in three places to check that the bearing is round and on size. Usually all of the connecting rod bearing internal diameters will be 'in tolerance,' which means within 0.006mm/0.0002in of each other.

The measurements are recorded and checked against the factory specification. The crankshaft journal sizes are then subtracted from the actual bearing internal diameters: the difference between the two sizes is the bearing clearance.

If the desired clearance is not obtainable the problem lies with either the crankshaft journal size or the size of the bearing bore after assembly. Double check all sizes to find out which components are not to size. Because of manufacturing tolerances both may be wrong and generally the crankshaft journal can be 'eased' to suit the bearing sizes.

The connecting rods are now fully refurbished so that, if the crankshaft is to be reground, the connecting rod bearing sizes are known and the crankshaft journals can be ground to suit to give optimum bearing clearances. If the connecting rods are resized so that the bearing has the maximum amount of crush, the size of the bearing bore will usually be on or near the minimum size. The crankshaft may also have to be ground to the minimum size or 'bottom size' within the factory tolerances.

Optimum connecting rod bearing clearances

All Pinto engines require connecting rod bearing clearances of 0.05mm/0.002 minimum to 0.06mm/0.0023in maximum.

CRANKSHAFT

Pinto crankshafts are well machined and finished when they are manufactured and note that the journals are 'roll radiused' which increases the crank's strength. The Pinto crankshaft is also quite substantial: failures are few and far between. A crankshaft with standard journal sizes is best. Ensure that all journals are polished to a mirror finish. The ideal crankshaft for high-performance use is one that is crack-free, standard in journal size and needing only a journal polish.

If a crankshaft has been damaged through bearing failure its journals will almost always need to be reground. When contemplating regrinding a crankshaft, the diameter size of the connecting rod journal is more critical than the main bearing

Roll radii which strenghten crankshaft.

SHORT BLOCK REBUILD

Pinto crankshaft.

size, simply because the mains are always larger in diameter and are, as a consequence, stronger. It is acceptable to regrind main bearings to 1.5mm/0.060in undersize but consider 0.75mm/0.030in the *maximum* undersize when regrinding connecting rod journals for anything other than a stock engine.

CRANKSHAFT CHECKING
Crack testing

All the external surfaces of the crankshaft *must* be crack tested with special attention being paid to each journal's radii. Crankshafts crack in many places but, by far the most common, is in the corner radius of a journal. Any crankshaft that has a crack, *must* be replaced. Note that normal crack testing equipment, as used by engine machine shops, can only check the surface of the crankshaft for cracks.

The crankshaft is magnetized using an electromagnet and a mixture of 'Magnaflux' powder and kerosene is sprayed over it. If there is a crack the powder will congregate along the crack and highlight the crack. The operator must be skilled in the use of this equipment as a crack can be missed by the inexperienced. The crankshaft is demagnetized after the crack test procedure.

The fact that a crankshaft passes the crack test does not mean that it is totally crack-free. The type of procedure previously described will not pick up an internal crack. The only way to check a crankshaft completely is to have it x-rayed. Cracks can start inside the crankshaft from the drilled oilways.

Any crack testing procedure can only show that, at the time of the test, the crankshaft was crack-free. A crankshaft can develop cracks shortly after it is put back in service. Having a crankshaft pass a crack test is not a guarantee that it will not crack and break in service.

Checking straightness

The crankshaft *must* be straightness tested. If you can, do it between centres of a lathe or in a crankshaft grinder, although this check can also be carried out using the engine block with only the front and rear bearing inserts fitted. A dial indicator and magnetic stand can be used to check the 'run out' of each main journal. Pinto engines have central thrusts and these should be in place (unoiled) when the crankshaft is turned to prevent possible damage to the crankshaft thrust surfaces through block contact.

Measuring crankshaft runout (straightness).

SPEEDPRO SERIES

Crankshafts *must* be dead straight (zero runout) for high-performance use, and most are anyway. If necessary, straightness can be restored by remachining the journals.

Crankshaft detailing

The oil holes *must* be radiused where they meet the crankshaft journal surfaces (main bearing and connecting rod bearing journals). Frequently, the edge formed by the hole as it breaks out on to the journal surface is razor sharp. If the crankshaft is to be reground, the oil holes should be radiused *before* the regrind to avoid the possibility of journal damage, though you'll need to check that they are still okay after the regrind. A high speed die grinder, with a 4mm/3/16in diameter mounted point fitted, is the ideal tool to do this job. **Warning:** wear protective gloves and safety goggles when using a grinder. **Caution:** care is needed when grinding to avoid slipping and damaging nearby surfaces (this is less critical if the crankshaft is going to be reground). Work slowly with minimal side pressure on the stone to avoid slipping or running over the journal.

Radiusing oilways.

If the journals are on size and the crankshaft does not need to be reground, the oil holes will have to be ground *very* carefully to avoid slipping with the high speed grinder and marking the journal surface. One slip with the grinder and the journal surface will be marked. *Absolute care is vital*.

Clean all unmachined surfaces with a rotary wire brush to remove all dirt and grime. If the crankshaft has been cleaned in a hot tank, there will be very little to clean off. **Warning:** wear protective gloves and safety goggles when working with a rotary brush.

To prevent damage to journal surfaces during crankshaft cleaning, the journals can be 'masked' by placing a pair of old bearing inserts on to the journal surface and then wrapping electrical tape around the outside of the inserts to hold the inserts in place. If you slip with the rotary wire brush it will run across the tape and the back of a bearing insert.

Scour all crankshaft oilways with a square-ended mild steel rod to remove all traces of dirt and grime. An engine that has not had regular oil changes will most likely have a considerable build-up of dirt and grime in the oilways. The mild steel rod will not scratch the inside of the crankshaft's oilways.

Most crankshafts have rough edges on them and they can be quite sharp. Remove all casting flash (particularly around holes in the throws of the crankshaft) and partlines using a high-speed grinder with a mounted point stone. **Warning!** Wear protective gloves and safety goggles when working with a grinder.

Clean out the threaded holes in the rear crankshaft flange with a tap. The tap recuts the threads so that

Typical moulding partline which should be smoothed off with a grinding tool.

the crankshaft flywheel bolts can be wound in easily and torqued without interference from threads that are distorted or dirty.

There is no substitute for a crankshaft that is dead straight, has perfectly round journals, precise clearances on each and every bearing, well radiused oil holes, a flawless surface finish on each of the journal surfaces and the casting flash smoothed over by a high-speed grinder and mounted point grinding stone. It should be virtually impossible to cut yourself after the crankshaft has been detailed. Note that all edges which have been ground should be further smoothed by hand using 120 grit wet-and-dry paper or cloth tape.

SHORT BLOCK REBUILD

Measuring journals

The crankshaft journals are now measured with a micrometer to check that they are on size. Measuring like this assumes that the journal was originally round and that any wear has made the journal oval.

The optimum main bearing journal sizes are ascertained by measuring the bearing insert bores when in the block and subtracting the correct running clearance (which should be between the middle and top of the engine manufacturer's recommended tolerances), which then gives the actual size required for the main journal. The same applies to the connecting rod journals. The bearing bores of the connecting rods are measured and the clearance added which then gives the correct diameter for the journals. Note that individual journals can be ground to different sizes to give optimum clearances on all bearings.

CRANKSHAFT REGRINDING

When regrinding a crankshaft the connecting rod journal diameters are always ground first if they are being reground. The reason for this is that there can be slight flexing of the crankshaft during regrinding and, if this happens when the mains have already been finished, the crankshaft will not run true and the mains may have to be reground a further undersize to correct the problem. Crankshaft journal surfaces need to have as near a mirror-finish as possible.

If the connecting rods have not been refurbished and measured before the crankshaft is reground, the optimum journal sizes are not known. This is why the connecting rod and the main bearing inserts are fitted and their inside diameters measured before the crankshaft is reground. If the crankshaft is 'on size' and the connecting rods resized, the bearing bore diameter with the inserts fitted may well be reduced as a consequence and the crankshaft may have to be reworked. In such a situation the crankshaft journal

Measuring main and big end journals.

is 'eased' to achieve the desired clearance. Tight bearing clearances are not desirable. If the bearing tunnel diameters are known before the crankshaft is reground there are definite sizes available to grind the crankshaft to. This is the correct way of obtaining the correct running bearing clearance and it is a logical sequence requiring the least work.

'CHECK-FITTING' CONNECTING RODS TO CRANKSHAFT

With the crankshaft reground or, if the journals did not require regrinding and were polished only, the connecting rods are 'check fitted' to the crankshaft to check their freedom of rotation.

If there is anything wrong, such as insufficient bearing clearance or connecting rod side clearance, it will be found now and can be corrected at this point.

With the crankshaft lying on a bench, each connecting rod is fitted to the correct journal and torqued up just as it would be when assembled in the block. Do not apply engine oil to the journals and the bearing surfaces at this point. *Caution*: before the connecting rod bolts are torqued up, insert an appropriately sized feeler gauge between the connecting rod and crank to prevent any sideways twisting movement of the connecting rod.

Note that by placing a correctly-sized feeler gauge between the connecting rod and crankshaft, the side clearance is also ascertained. The side clearance must be within Ford's tolerance limits; not too tight and not too loose. Seldom will the connecting rod side clearances be less than 0.104mm/0.004in or more than 0.15mm/0.006in. Check all side clearances are to the manufacturer's specifications.

SPEEDPRO SERIES

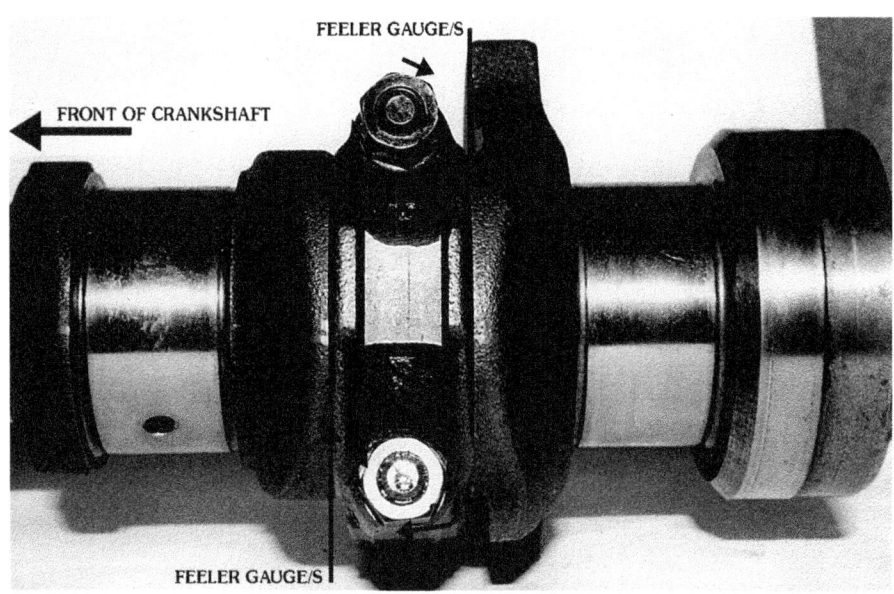

Feeler gauges should be positioned as shown to prevent bearing damage when the connecting rod bolts are tightened.

Connecting rod check-fitted to crankshaft to test freedom of rotation.

Caution: do not rotate the connecting rods on the crankshaft journals until the nuts are torqued up. The reason for this is that, until the nuts are torqued up, the two inserts have not been forced to take the shape of the connecting rod tunnel and the actual surfaces of the bearing inserts can be damaged.

Caution: check that the socket used to tighten the nuts does not contact the connecting rod cap. If the socket does contact the cap it can become misaligned or the nut can give a false torque reading.

With the connecting rods fitted, the crankshaft is lifted up so that it is vertical (resting on the rear flange), or the nose is clamped in a vice with protective shields fitted between the vice jaws and the nose.

The connecting rods are each then checked for freedom of rotation for the full 360 degrees. By turning the connecting rods one at a time any binding will be felt. If there is binding, remove that particular connecting rod, refit the cap and then re-measure the bearing insert bore diameter in three places and especially across the bore where there are 'scuff' marks. This is to check the overall roundness of the bearing bore. **Caution:** use a feeler gauge between the connecting rod and crankshaft during all assembling and dismantling to avoid any possibility of damage to the bearing shell inserts.

If there is binding there will be some scuffing of the bearing insert surface but it can be very difficult to see at times. Use a magnifying glass if necessary. Bearing inserts can have high spots at times but this is unusual. If there is a high spot it can be eased with a three-corner scraper but *never* with wet-and-dry abrasive paper. Check to make sure that it is a high spot and that there is not a general overall lack of bearing clearance or debris behind the inserts. Check the bearing bore size of the connecting rod and for misalignment of the cap as this can be caused by the socket becoming jammed against the connecting rod cap.

This is a dummy assembly designed to check the actual fit of all of the connecting rods on the crankshaft. All of the sizing was established by direct measurement using micrometers; the fitting of the connecting rods is done to double check working clearances.

With this procedure carried out and the bearing clearances proved correct, or found to be incorrect and the situation remedied, the connecting rod to crankshaft check fitting is complete.

BUILDING THE SHORT BLOCK

Freeze plugs have been known to come out of cylinder blocks and cylinder heads. Any competition engine needs to have some form of retention across all of the freeze plugs so that if one comes lose at least it will not fall out.

When the crankshaft is being installed it *must* be very carefully

46

SHORT BLOCK REBUILD

lowered into the oiled bearings in the block and *must not* be rotated. The main bearing caps (oiled bearing shells and thrust washers included) are then fitted to the block and snapped in, the bolts fitted and torqued to the correct tension. Only now can the crankshaft be turned because the shell bearings are not truly round until the bolts have been torqued up.

When installed the crankshaft must turn freely (turn for one revolution – or very near it when spun by a flick of the wrist). Check the crankshaft endfloat (endplay/lash) and the connecting rod side clearances with feeler gauges.

The pistons and connecting rods can now be fitted to the block (**Caution!** Cover the rod bolts with petrol pipe to prevent damage to the crank journals), the con-rod caps fitted and the nuts (or bolts on some rods) torqued to the specified tension. **Caution!** When tightening the connecting rod nuts feeler gauges should be inserted between the side of the connecting rod and the crankshaft to prevent the twisting forces from causing damage to the bearing shells.

Two 4mm cap screws in drilled and tapped holes hold these two washers over the edge of the freeze plug bore.

When the short block assembly is completely assembled, the engine should be able to be turned by hand very easily.

www.velocebooks.com / www.veloce.co.uk
All current books • New book news • Special offers • Gift vouchers

Chapter 5
Cylinder head

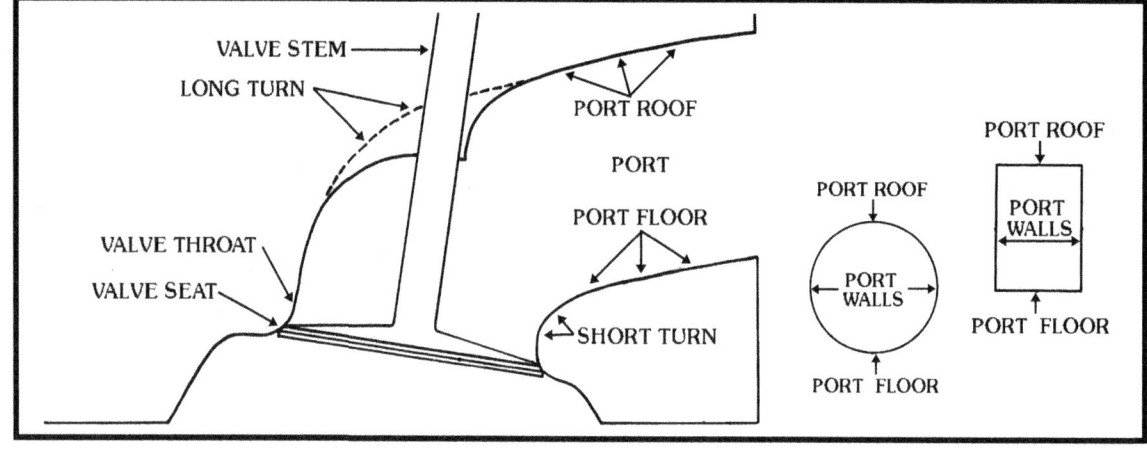

Port terminology.

The cylinder heads for the three engines (1600/1800/2000) are similar in layout with only very subtle differences between them. The differences are the size of the valves, the length of the valves, the size of the ports, the angle of some inlet ports, inlet port floor radius, valve throat contour, timing of the camshafts and the sizes of combustion chambers.

All sorts of head modifications have been tried over the years in an effort to extract more power from Pinto engines. The more radical designs/ modifications have been costly and, in the final analysis, the improvement in overall power has been disappointing. Because it is the most cost-effective way of obtaining extra power, the cylinder head porting described in this book deals only with what can be done to a standard head without adding material (by welding, brazing or two pot mix epoxy resin substances) and removing only what is necessary to improve the shape of the existing ports (inlet and exhaust). That said, a considerable amount of reworking is still necessary during the modification process.

The fact cannot be ignored that many cylinder heads are altered using epoxy resin, with good results being achieved. The epoxy resin is used to fill the lower part of the inlet port right up to within 1/4in/6mm of the valve seat. A nice shape results. The maximum power may not be any better but the low end and mid-range pickup

CYLINDER HEAD

is almost always improved (that's the power or 'push' out of a turn).

Going to the extremes of raising the port higher than shown means going through the water jacket (and the very likely prospect of continual water leaks) and this is not recommended even though the idea is right.

The combustion chamber shapes are similar for the three engines, but their sizes do vary slightly. The depths of the combustion chambers also vary, and it this feature of the Pinto design that necessitates slightly different length valves for the different capacities in order to maintain good rocker geometry.

The camshafts are all very similar and will interchange.

There are two basic inlet port designs. The first is the type as found on nearly all Pinto engines from start to finish of production and the second is the Sierra IS head which is only found on that engine (1986 on).

The standard cylinder head inlet ports are quite large and the overall diameters of the ports range between 35mm/1.375in and 38mm/1.495in. The 2000cc engines have the largest inlet ports (vary between 36mm/1.415in and 38mm/1.495in), the 1600cc engines have 35mm/1.375in to 35.5mm/1.395in and the 1800cc engines have 35.5mm/1.395in to 36.5mm/1.435in.

VALVES

The valves for all three engines have different head diameters and different lengths but note that, with the addition of 'lash caps' (which increase the effective length of a valve), any valve from a larger valved head (Group 1 for instance) can be installed into what was originally a smaller valved head if necessary.

The standard (and Group 1) valves all have triple keeper grooves which, in conjunction with the keeper to valve stem clearance, allows the valve to rotate freely, promoting even valve and seat wear. This is an excellent design for road-going engines but it is not ideal for high-performance engines turning over 6700rpm on a continuous basis. Over this engine speed (anything over 6700rpm as a normal gear change point) the grooves wear very quickly and then there is excessive lateral movement between the keepers and the valve stem. Single groove valve stems and matching keepers are recommended for sustained high rpm operation (anything over 7000rpm as a normal gearchange point).

The 1600cc engine's valve diameters are 42.1mm/1.672in inlet and 34.5mm/1.357in exhaust. These valve sizes are sufficient for a high-performance four-cylinder engine of this size. The inlet, especially, is large for an engine of this capacity.

The 1800cc engine's valve diameters are 41.3mm/1.625in inlet and 34.2mm/1.345in exhaust. These valve sizes are sufficient for a high-performance four-cylinder engine of this size.

The 2000cc engine's valve diameters are 42.10/1.657in inlet and 36.2mm/1.425in exhaust. These valve sizes are reasonable, but if both inlet and exhaust valves are replaced by larger valves, engine torque is increased, along with top end (5500rpm and up) power.

The standard two-piece valves are only reliable for 6700rpm operation. They are quite heavy compared to stainless steel one piece competition valves which are 'upset forged' and are usually made from a material called EN24B.

These stainless steel competition/racing valves are not expensive and represent very good value for money because valve failure at high rpm (8000rpm plus) is not common once they have been fitted correctly to a cylinder head. They are the same length as Group 1 valves (Group 1 was a European racing class for which Ford homologated a range of parts), so lash caps may be necessary to correct rocker geometry, making them ideal for high lift camshafts with smaller than standard base circle diameters.

Valve lengths are -

Inlet
1600 std - 113.13mm/4.453in.
1800 std - 112.06mm/4.412in.
2000 std - 110.55mm/4.325in.
Group 1 - 112.50mm/4.429in.

Exhaust
1600 std - 112.69mm/4.436in.
1800 std - 111.66mm/4.396in.
2000 std - 110.55mm/4.352in.
Group 1 - 110.50mm/4.350in.

The 2000cc engine has a deeper combustion chamber (shortest valve stems) than the 1800cc and 1600cc engines; the 1800cc engine has a deeper chamber than the 1600cc engine (longest standard engine valve stems). The variations in valve lengths place the tops of the valve stems in approximately the same place on all cylinder heads so that the geometry of the rockers is correct.

The standard position of the top of the valve stem is 120mm from the cylinder head gasket surface, but this measurement is only good if the cylinder head has *not* been planed. Also, on used engines, there is likely to be some variation in the valve heights as their seats might have been reground, effectively increasing the height of the valve/s as they are deeper seated in the head. There is some latitude in the fitted height of the valve stem top, but consider plus

SPEEDPRO SERIES

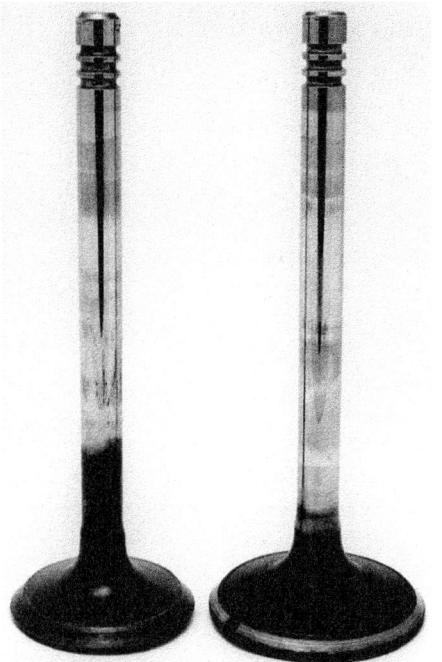

Above, left: Standard exhaust valve (left) and inlet (right). Above, right: Original Group 1 exhaust valve (left) and inlet valve (right). Both standard and Group 1 valves have triple keeper grooves.

Aftermarket Group 1-sized replacement valves. Note waisted valve stems on both valves and triple grooves for standard valve keepers.

The benefit of larger valves

Fitting larger valves increases the area, or 'window,' through which gas can flow at any given amount of lift (from the seat). A seemingly small increase in the diameter of a valve does, in fact, make a lot of difference to gas flow and will definitely improve the volumetric efficiency of the larger Pinto engines (**Caution!** Larger valves than standard size cannot be used in the 1600 as they'll foul the cylinders).

Note that the effective gain from fitting larger valves applies only up to the amount of lift where the area of the 'window' equals the port/throat combination's maximum gasflow potential. Beyond this lift point, the smaller standard valve will flow an equal amount of air until the limit of port/throat gasflow is reached.

For the 1800 and 2000 heads the Group 1 valve sizes are quite appropriate as they allow the inlet and exhaust valve throat passages to be well ported, yet still retain a suitable amount of material between the inlet and exhaust valve throats.

Group 1-type valves (1800/2000 engines)

These larger valves have long been available for the 2000 engine and are

or minus 0.50mm/0.020in to be the maximum tolerance for absolute valve train reliability. (Ford list a standard tolerance of 1.0mm /0.040in).

Note that for correct rocker geometry when using a non-standard camshaft the actual length of the valve is largely irrelevant as a 'lash cap' can be used to compensate for shortness.

42.1mm In.
44.4mm In.
45.4mm In.
34.5mm Ex.
36mm Ex.
38mm Ex.

Comparison of valve 'windows' (effectively the circumference of the seat rolled out to form a strip).

CYLINDER HEAD

Standard 2000cc valves in standard combustion chamber and the 3.5mm/0.137in gap between the valves.

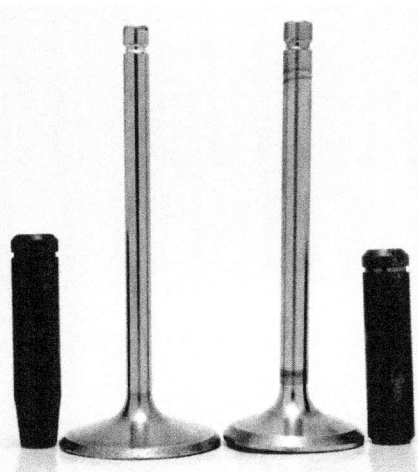

Holbay single groove inlet valve and valve guide (left) and exhaust valve and valve guide (right).

44.4mm/1.750in in diameter (inlet) and 38mm/1.495in in diameter (exhaust). Useful gasflow gains are made by fitting these valves to the 1800/2000cc head, provided the valve throats and seats are enlarged correctly to suit the size of the valve. These valves also have longer stems to restore the valve train geometry to standard when they're used with the Group1 camshaft (for which they were designed).

The original Group 1 valves have triple keeper grooves which makes them compatible with the standard keepers and standard valve retainers. However, with triple groove valves, revs must be limited to 6700rpm as anything above this (on a continuous basis) tends to wear the grooves out and will eventually result in failure. The solution to this problem is to fit single groove valves and keepers which do not allow the valve to rotate at all. These aftermarket single groove inlet and exhaust valves and keepers are available from specialist parts suppliers such as Burton Power, Vulcan Engineering and Holbay Engineering. Holbay Engineering have only ever made single groove valves and keepers in their line of racing valves. Their keepers are turned out of high tensile round bar as are their matching valve spring retainers. All valve head sizes and valve stem lengths are available in single groove design from these suppliers.

An engine machine shop (engine reconditioner) can accurately enlarge the valve seats to suit the larger valves using conventional seat cutting equipment. The new seat sizes are 44.4mm/1.750in (inlet) and 38.0mm/1.495in (exhaust). The new seat width is 1.25mm/0.050in (inlet) and 2.0mm or 0.080in (exhaust).

The top edge of each seat is given a 1.0mm/0.040in wide 'top cut' at 30 degrees and the 'inner cut' is flared from the original port throat to the new valve seat. Otherwise the port remains completely standard.

On the standard 2000cc cylinder head with standard valves fitted there is 3.5mm/0.137in (seated) between the valve heads at their closest point. When Group 1 valves are fitted, the gap is reduced to 1.25mm/0.050in (with standard valve guide centres) which is approaching the desirable proximity limit.

Group 1 inlet valves are 44.4mm/1.750in in diameter and, when the throats are cleaned up, they should *not* be made larger than 38.5mm/1.515in (39.0mm/1.535in as an *absolute maximum*).

Group 2 inlet valves (2000 engines)

The largest valves usually fitted to the 2000 Pinto cylinder head (using the standard valve centres) are Group 2 inlets which have head diameters of 45.4mm/1.785in (the Group 1 valve is used for exhaust). This combination leaves just 0.75mm/0.030in (seated) between the heads of the valves. This is very close and, if the valve throats

SPEEDPRO SERIES

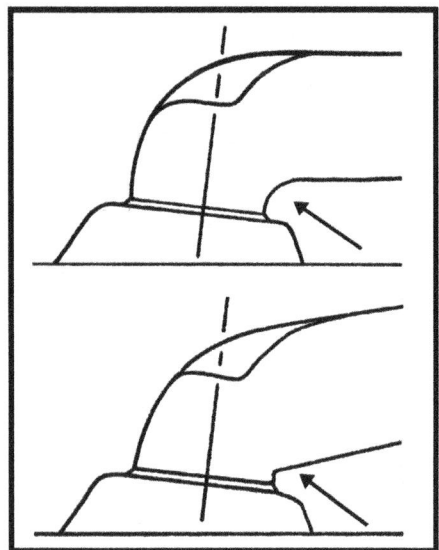

Top - Sierra IS head's port shape. Below - shape of standard Pinto inlet port near valve head.

Group 1 valves in modified combustion chamber of 2000cc cylinder head. The gap between the valves is 1.25mm/0.050in.

were opened out to the absolute maximum, it would be too close because there would be insufficient head material between the two valve seats. The length of the Group 2 valve is the same as that of the Group 1.

This valve combination creates some risk of cracking between the inlet and exhaust valve seats but, if the valve throats are ported in a particular manner, the prospect of cracking is greatly reduced. The method involves not boring the inlet valve throats too large, so that sufficient material thickness remains between the two valve seats and, although the actual valve seats are very close, there's a considerable amount of material still there. The *maximum* throat size recommended for the largest-sized inlet valve is 39.0mm/1.535in. The largest readily available exhaust valve diameter of 38mm should have a *maximum* valve throat size of 34.5mm/1.360in.

Caution! Note that if the amount of material left between the throats after modification is insufficient, the head will crack (sooner or later) between the inlet and exhaust valve seats and render the cylinder head useless. When these large valves are fitted the valve throats *must* be ported in such a way that a mean thickness of *not less* than 5.1mm/0.200in remains between the valve throats. Obviously, the seat-to-seat dimension is much closer than this, but the material thickness below the seats is to the prescribed thickness as close to the seats as is possible.

Sierra IS head/valves

The later Sierra IS engine, which features fuel-injection, has standard sized inlet and exhaust valve diameters of 42mm and 36mm respectively. A significant design improvement in this cylinder head is the way in which the inlet port turns into the valve throat on the short side of the port: it has a radius, not the almost sharp-edged turn found on other Pinto heads (see diagrams). The IS cylinder head casting is much smoother in shape, although it features an enlarged cut-out to accommodate the fuel-injection nozzle which, if the engine is converted to carburettors, is redundant. The cut away area for the injectors is not that large and can, essentially, be ignored.

The IS unit is a good cylinder head to modify and fit to any modified 2000 Pinto engine. The inlet port floor is not angled down and, because of this feature, a reasonably well-shaped radius can be made on the short side of the turn into the inlet valve throat. The standard port diameter is already quite large at 38.0mm/1.495in diameter. Don't enlarge the port from this size, leave it as cast.

For any high-performance 2000/2100 Pinto engine the IS component is an ideal cylinder head, needing only slight cleaning up of the

CYLINDER HEAD

inlet throats. The large size Group 1 valves can be fitted to these heads with the appropriate valve seat re-working as per any other Pinto cylinder head. The exhaust ports need to be opened out (as described later in this chapter) because they are the same as found on any other Pinto head.

Valve size summary

In fact all Pinto engines will run very well with their standard valve sizes as long as valve throats are modified.

1600 engines retain the standard valve sizes but one-piece 'stainless' valves are recommended. The cylinder bore size (81.3mm to 82.8mm) limits the valve head sizes that can be used in this engine so don't use valves larger than those of the 2000 Pinto (42mm inlet, 36mm exhaust). Expect to have to use lash caps to optimise rocker geometry.

1800 engines can either retain the standard valve sizes or have Group 1 valves fitted. The standard valve sizes are suitable for modified engines but the larger group 1 valves will fit without fouling the cylinder bore (86.2mm to 87.7mm diameter) and do allow better cylinder filling. With the latter, expect to have to use lash caps to optimise rocker geometry.

2000cc engines can either retain the standard size valves or be fitted with Group 1 valves. Depending on the camshaft used, lash caps will have to be used to optimise rocker geometry.

2100cc engines (2000cc engines bored out to 93mm) benefit from the use of Group 1 inlet and exhaust valves, but will still run very well in high-performance applications with standard sized valves. Triple groove or single groove valve stems are available, base your decision on the rpm requirements of the engine.

Caution! For all engines and all

Standard 2000cc combustion chamber with valves removed so that the valve throats and guide bosses are clearly visible. The as-cast valve throat sizes of this cylinder head are approximately 36.0mm/1.415in for the inlets and 30.5mm/1.200in for the exhausts.

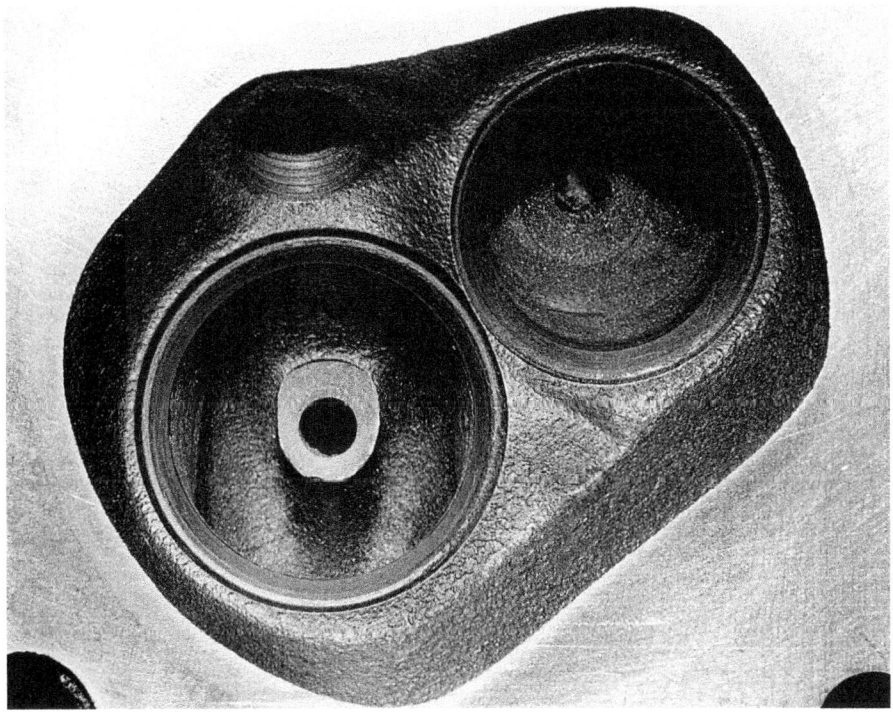

2000cc combustion chamber which has had the inlet throat only opened out to 38.5mm/1.515in.

SPEEDPRO SERIES

valve sizes/types using up to 6700rpm you can use triple groove valve stems but, for engines (mainly competition) which will use over 6700rpm on a continuous basis, you *must* use single groove valve stems and appropriate keepers. If triple groove valve stems are used in an engine occasionally revved to 7000rpm, the increase in wear of the keeper grooves will be negligible.

VALVE THROAT SIZE

The valve throats of all Pinto cylinder heads should be opened out to the optimum size for the valves used. The following table gives the optimum sizes to which valve throats can be opened out with absolute reliability of operation and with a considerable overall improvement in gasflow over standard –

Exhaust valves	Open throat to
34.2mm valves	30.5mm
34.5mm valves	31.0mm
36.0mm valves	32.5mm
38.0mm valves	34.5mm
Inlet valves	**Open throat to**
41.5mm valves	38.0mm
42.0mm valves	38.5mm
44.4mm valves	39.0mm
45.4mm valves	39.0mm

If standard valve sizes are going to be retained, the inlet and exhaust ports should *not* be opened out, but the valve throats *must* be modified (this involves reworking the inlet and exhaust throat and bowl areas only).

INLET PORT SIZE (STD SIZE VALVES)

The 36 to 38mm port diameter of all 2000 cylinder heads is optimally sized just as it comes from Ford and does not need to be opened out at

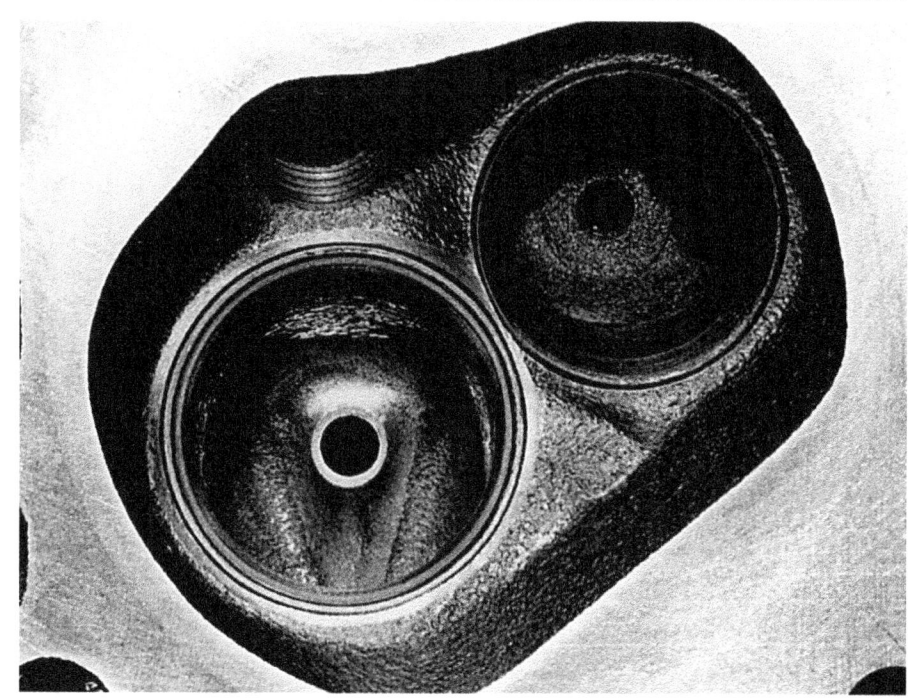

Completed modifications to inlet port, valve throat, valve seat, valve seat ridge and valve guide boss. Valve seat is 1.3mm/0.050in wide.

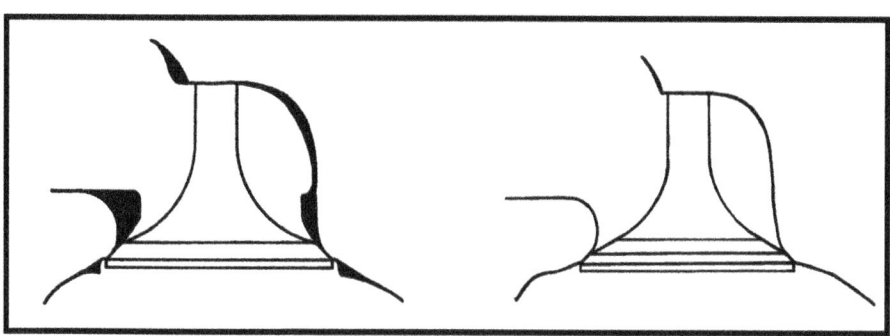

Black areas show material to be removed. The picture on the right shows how the valve head area of the port and combustion chamber should look after modification.

A selection of carbide tools and grinding stone used for porting work.

CYLINDER HEAD

all. In fact, it can be left in the as-cast state. The largest standard port size (38mm) is as large as the largest carburettor choke (38mm) that is normally used on a well modified 2000cc/2100cc engine. None of the Pinto cylinder heads, if fitted to their original block, really need to have their inlet ports opened out.

EXHAUST PORT SIZE (STD SIZE VALVES)

The exhaust ports of all the Pinto cylinder heads are very similar in size. When the standard-sized exhaust valves are retained, the ports do not need to be opened out. The ports do, however, need to be well blended (where the valve throat turns into the exhaust port proper) with the aim of

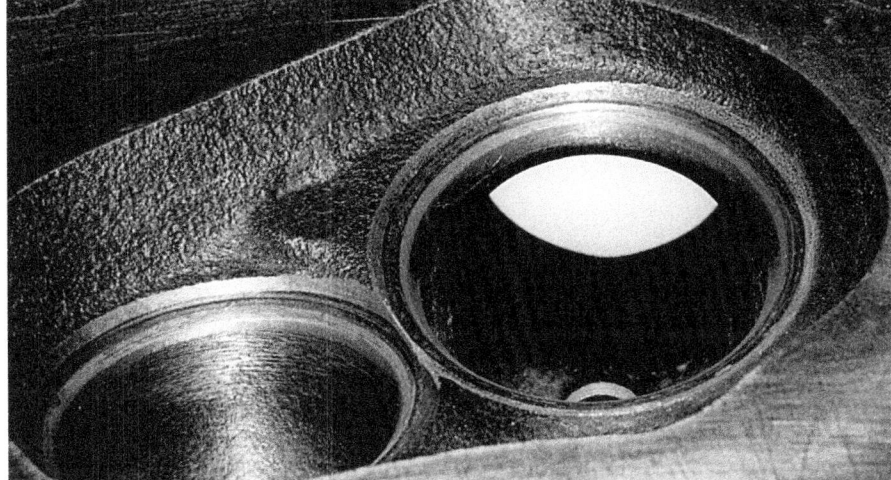

Top - Ridges (arrowed) left during original factory machining and standard inlet port (short) turn. Above - ridges have been removed inlet port turn modified.

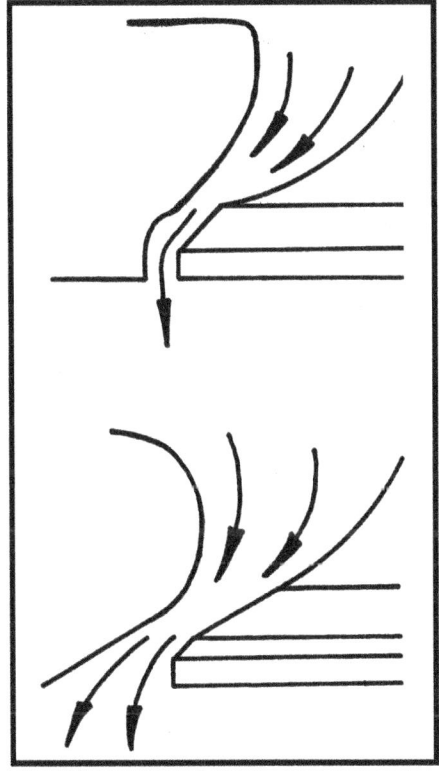

How gasflow improves at same valve lift when the valve is unmasked and the port reshaped.

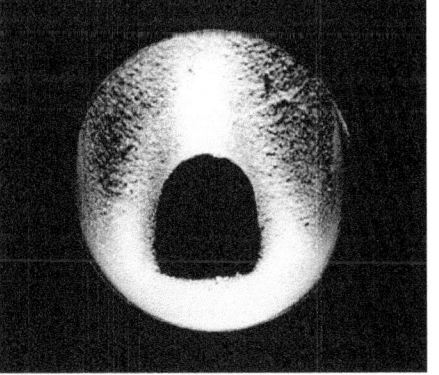 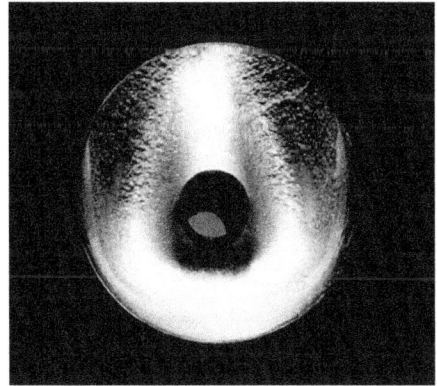

Standard valve guide boss (left) and modified valve guide boss (right).

SPEEDPRO SERIES

keeping the cross-sectional area of this part of the port at least equal to the rest of it.

In the interests of simplicity, yet still with a worthwhile increase in power, the modifications described are confined to the valve seats, valve throat, guide boss, port walls and the spot facing ridge in the combustion chamber.

VALVE THROAT & PORT MODIFICATIONS (STD SIZE VALVES)

The first stage of modifying any Pinto head is to open out the valve throats to the size appropriate to the valves. The valve seat is narrowed when the throat diameter is opened out and the throat area is blended back into the port, but *only* enough to make the transition smooth. The short turn from the valve throat into the port proper must be radiused, and the port floor turn flattened as much as possible. The ridge that is created by the spot facing of the top cut of the valve seat *must* be removed.

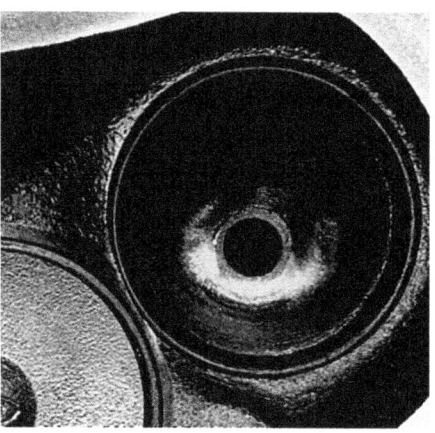

Modified valve guide boss. It has not been shortened.

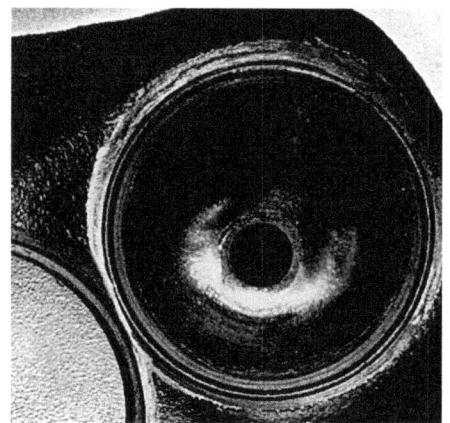

Completed modified exhaust valve seat, throat, port and unshrouded valve seat. The valve seat is 1.5mm/0.060in wide.

The example of porting shown in the accompanying photos was carried out on a standard 2000 cylinder head; the inlet ports of this particular head were found to be, on average, 36mm in diameter and not 38mm as many are.

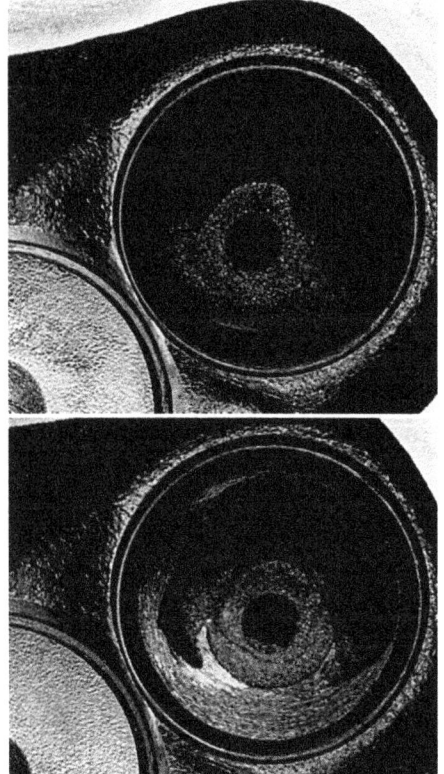

Top - standard exhaust port. Above - modified exhaust port with throat opened out to 32.5mm/1.28in diameter.

Completed combustion chamber work with the spot facing ridges removed to unshroud the valves. The chamber shape is unaltered.

CYLINDER HEAD

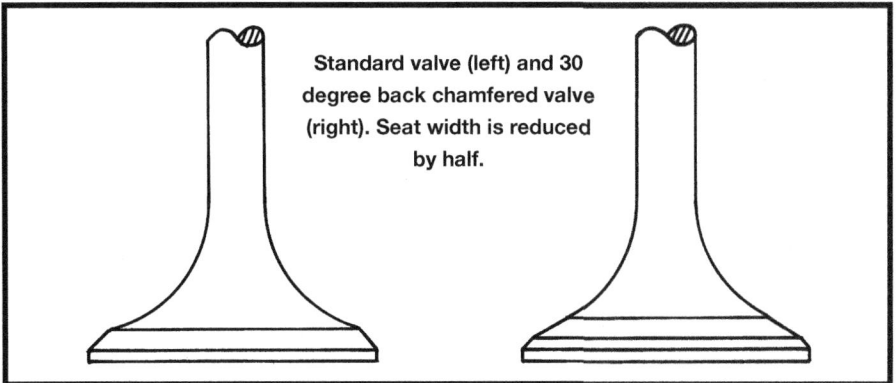

Standard valve (left) and 30 degree back chamfered valve (right). Seat width is reduced by half.

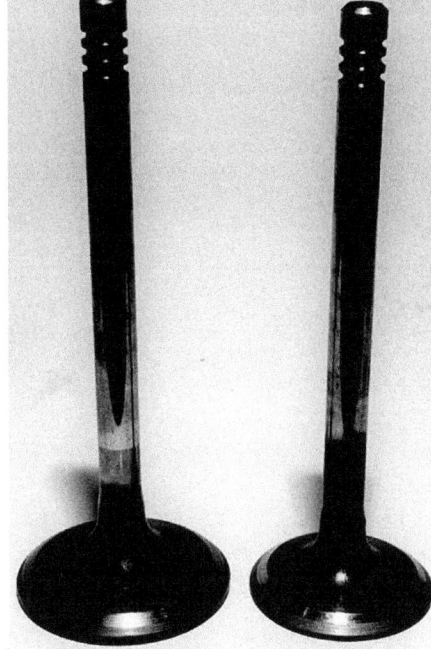

Inlet valve (left) and exhaust valve (right). Both valves have been back chamfered 30 degrees.

Close-up of back chamfer. Note that the back chamfer intersects the valve seat, reducing its width.

Valve unmasking

The valve seat is itself in a machined recess or 'spot facing' in the cylinder head which creates a ridge (or wall) of varying height directly adjacent to the margin of the valve when the valve is seated. This ridge, which is usually about 1.0mm/0.040in high, *must* be removed (use a small high-speed grinder). If the wall is not removed, there will not be an effective gasflow into the cylinder until the valve is at least 1.0mm/0.040in off its seat because of the ridge's masking effect. Protect the valve seat by using an old valve which has had its margin removed. This way, if the grinding wheel veers from the work area, it will only run over the old valve and not the valve seat.

Inlet ports

With the inlet valve throat opened out to the prescribed size and blended back into the port as shown in the photograph, the valve seat is then narrowed down so that its width is reduced to 1.3mm/0.055in. This process can be carried out by an engine machine shop (engine reconditioner) with valve seat grinding equipment. The standard valve seats could be anything up to 3.5mm/0.140in wide depending on how well they have been ground. The outer diameter of the valve seat *must* be the same as the valve head diameter.

The floor of the inlet port and the turn into the valve throat must be flattened off to a degree. The turn in, as originally machined, is sharp-edged and must be radiused as much as possible, though the scope for making a nice radius is very limited.

Exhaust ports

These are essentially the same for all Pinto head castings with the exception of valve size differences and corresponding valve throat sizes. The standard port passageway does not have a particularly smooth contour and the cross-sectional area is much less than it could be. The first consideration is the valve size because this determines the diameter that the valve throat is to be bored out to.

As an example, the 2000 (inc IS) cylinder head has a 36.2mm diameter exhaust valve and the standard 'as-cast/bored' valve throat is approximately 30.5mm at the smallest point. This is taken out to 32.5mm.

The valve guide boss protrusion into the port is not removed, but it is streamlined to such an extent that no flow losses are incurred by its retention. The guide boss length is left intact to improve the guide's service life and reliability. The port passage may look a bit larger and less cluttered without the guide boss but, in practice, there's very little advantage in gasflow terms. The recommendation is *not* to remove the valve guide boss.

The standard inlet and exhaust valves, in addition to being reground, should have a chamfer ground on their backs. If the valves are new (and they should be), they must be machined to give a back chamfer of 30 degrees. This operation is done on a lathe fitted with a tungsten carbide turning tool or on the valve regrinding equipment of an engine machine shop.

The back chamfer will vary in size according to the shape of the back of the valve. The actual contact width of the valve seat for all inlet valves must be 1.3mm/0.050in. The exhaust valves' seat contact width is 1.5mm/0.060in and this means that the inner edge of the valve seat must be enlarged until the seat is the matching width. The outer edge of the valve seat must be of the same diameter as the valve.

LARGE VALVE MODIFICATIONS

Essentially, cylinder heads prepared for competition on, for example, a 2000 engine will feature 44.4mm-45.4mm inlet valves and 38mm exhaust valves. The overall modified shape of the throats remains similar to that previously described, but the cross-sectional areas *must* be increased.

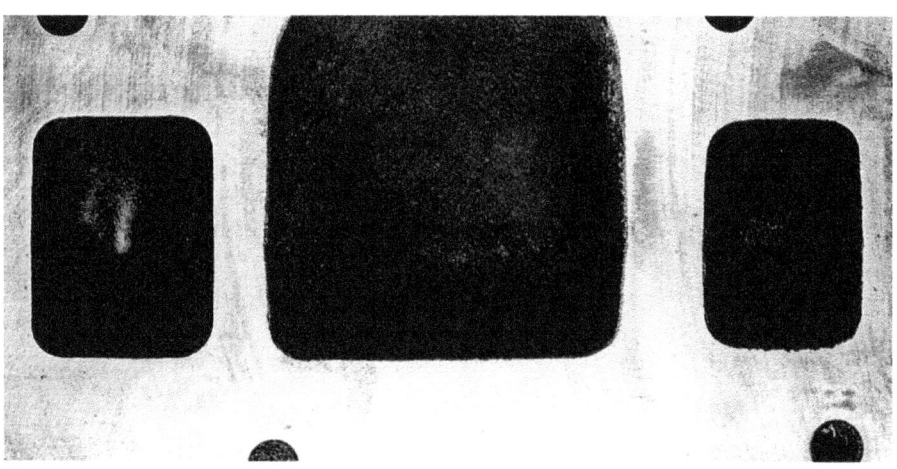
Completed modified 2000cc cylinder head exhaust port (left) which has been opened out to the standard gasket size. Completely standard port (right).

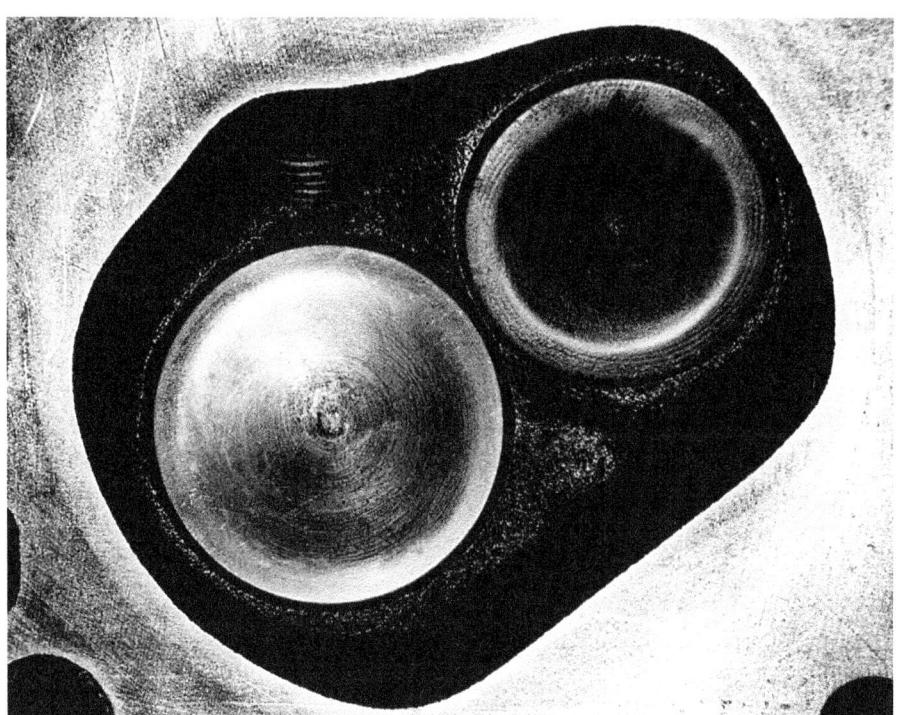
44.4mm inlet valve and 38mm exhaust valve fitted to a Pinto combustion chamber.

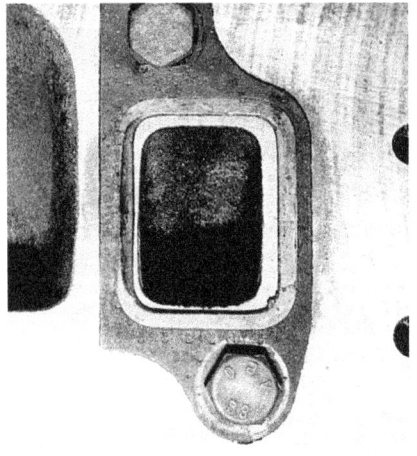
Exhaust manifold gasket bolted on to the side of the cylinder head (2000cc) so that it can be used as a template.

The inlet port can remain at 37 to 38mm in diameter and the guide bosses can be modified as previously described, but the valve throat diameter must be ground or bored out to 39mm/1.545in. The valve seats are recut to accommodate the new large valves. The valve seats must be 1.3mm/0.050in wide and a normal three angle valve seat is very suitable (45 degree seat/30 degree top cut/60 degree inner cut). The inner cut blends the 45 degree valve seat into the port throat and the top cut blends the 45

CYLINDER HEAD

degree valve seat into the combustion chamber.

The whole exhaust port passageway must be enlarged throughout. The valve throat is ground or bored out to 34.5mm/1.358in and the original valve seat machined to suit the new valve size. The port outlet is taken out to the size of the exhaust manifold gasket aperture (oversize – 1mm more height and width – exhaust manifold gaskets are available). The passageway is ground out quite extensively, but note that while the floor of the port should be cleaned up it should not have any serious amount of material removed from it, especially where the port runs into the valve throat area. The sides of the exhaust ports and the roof certainly have plenty of material which can be removed and, once modified, bear little resemblance to the originals.

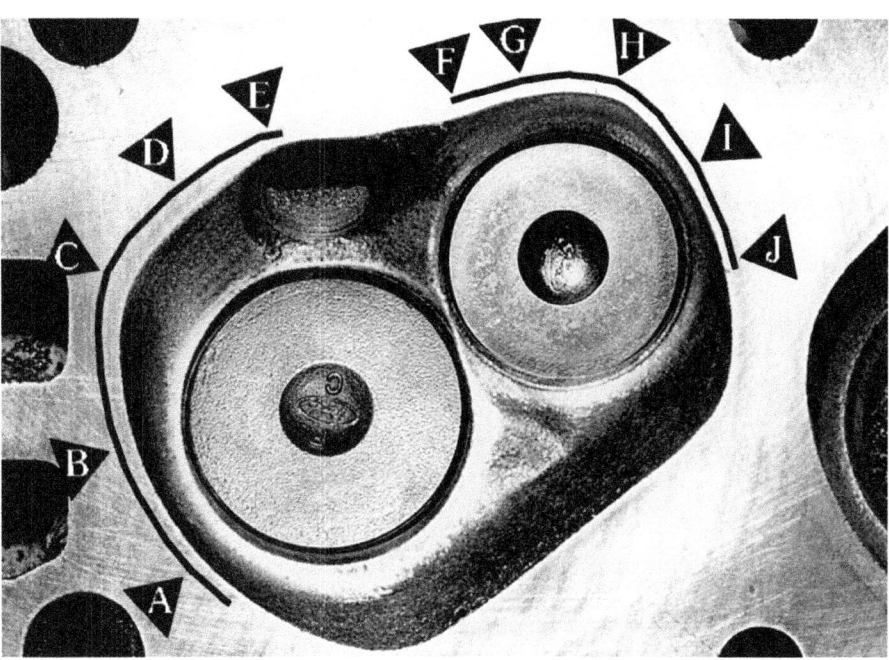

The areas of the combustion chamber to be reworked are shown here (see text for key). The amount of material that is removed does vary depending on the proximity of the gasket and bore wall.

Enlarging exhaust ports (large valves)

The modifications described here follow the methods universally used for Pinto engines. The valves are enlarged to the maximum size possible without altering the valve guide centre distance; the combustion chambers are unshrouded close to the cylinder head gasket (but not so close as to make the engine unreliable); the exhaust ports are enlarged considerably but the inlets are not; the valve throats are enlarged. These modifications produce very good results at a reasonable cost.

The exhaust port outlets at the side of the cylinder head (where the exhaust manifold fits) are opened out to the full size of the standard gasket using the gasket as a template. Oversized gaskets which allow an even larger port exit cross-sectional area are available, but are only required for competition engines which will frequently run to 7800rpm or more and, as a consequence, have large primary exhaust pipes (47.62mm/1.875in).

The standard 'as-cast' exhaust port exit measures on average 34.0mm/1.345in in height by 24.5mm/0.970in in width. This must be opened out to 35.0mm/1.380in high by 29.5mm/1.160in wide (the size of the standard gasket aperture). With an oversized exhaust gasket the size of the exhaust port exit needs to be 36mm/1.420in by 30.5mm/1.20in wide.

At the point 32mm/1.25in into the exhaust port from the exhaust manifold flange, the modified port should be 35.0mm/1.375in high and 1.150in/29.0mm wide (whichever type of gasket is used). This modification gives a considerable increase in exhaust port area because the standard 'as-cast' dimensions at this same point are 32.3mm/1.275in in height by 24.0mm/0.950in wide. Material is taken equally off the sides and roof (nearest the camshaft) of the port but *not* from the floor.

The side of the combustion chamber opposite the sparkplug is not touched during unshrouding work as it is far enough away from the valves not to cause any disturbance to gasflow. Adjacent to the inlet valve, the amount of material removed from the combustion chamber wall between 'A' and 'B' (see photo opposite) is approximately 1mm/0.040in, the amount of material removed between 'B' to 'C' increases from 1mm at 'B' to 3mm/0.120in at 'C,' between 'C' and 'D' the full 3mm is removed and then, between 'D' and 'E,' the amount of material removed reduces to 1mm. Adjacent to the exhaust valve the amount of material removed at 'F' is approximately 1mm/0.040in which increases to 2mm/0.080in at 'H,' it then reduces to approximately 1mm/0.040in at 'I' and continues at this amount to 'J.'

Essentially, the side of the combustion chamber immediately adjacent to the bore walls should be taken out as near to the bore size as it is possible to do. This means that the lines on the diagram above the lines A, B and C and then H, I and J should be more or less the same as the bore diameter.

Caution! The combustion chamber *must not* be opened out beyond the cylinder bore size.

Valve guides

It is not recommended that the valve guides be shortened (ground down or ground away) on these engines. They are not very long as standard let alone shorter than normal. The wear in the standard valve guides is also best restored by K-Lining the original worn guide bores as opposed to boring out the guide holes to take press fit replacement valve guides. K-Line valve guide inserts are extremely hard wearing, can be run with very tight clearances without seizing and can be replaced as many times as necessary for little cost.

Chapter 6
Compression ratio

Compression ratio (CR) is a measurement of how much an engine compresses the gas charge in each cylinder – 10:1, for example, means that the volume of the gas drawn into the cylinder is compressed until it occupies one-tenth of its original volume when the piston is at the top of its stroke. Basically, the higher the compression, conducive to the octane rating of the fuel being used, the more efficient the engine will burn the air/fuel mixture (go better). CR can be increased by using four basic methods or a combination of methods. The four methods are: cylinder head planing, fitting a thin head gasket, cylinder block planing and fitting raised top pistons. **Caution!** The amount of compression built into the engine must be matched to the octane rating of the fuel being used.

If an engine has been tested at 38 and then 36 degrees of total advance which is 'all in' by 3500 to 3700rpm, and the engine 'pinks'/'pings' under full throttle acceleration at and over this engine speed (assuming the air fuel mixture is correct), it is 'over compressed.' The CR will have to be reduced, a better quality fuel found or an octane booster used to increase the fuel's octane rating. There is no point in having so much compression that the engine cannot be accelerated using full throttle without 'pinking.' **Caution!** All Pinto engines were designed to run on leaded fuel. This means that they will 'suffer' from valve seat recession if straight unleaded fuel is used in an engine that is turning high rpm and under extreme loading. Pinto cylinder heads are made out of relatively 'soft cast iron'. High-performance Pinto engines burning unleaded fuel must use an additive (Millars, Red Line, Superblend, for example) in the fuel to prevent valve seat recession. Monitor just how well the additive you end up using in your engine is protecting your engine's valve seats by checking the exhaust valve clearance frequently. If the valve clearance is continually closing up, the valve seats are recessing.

HEAD PLANING
This process is normally carried out to increase the compression ratio of the engine or to true up the gasket surface of the head to prevent gasket problems. The removal of

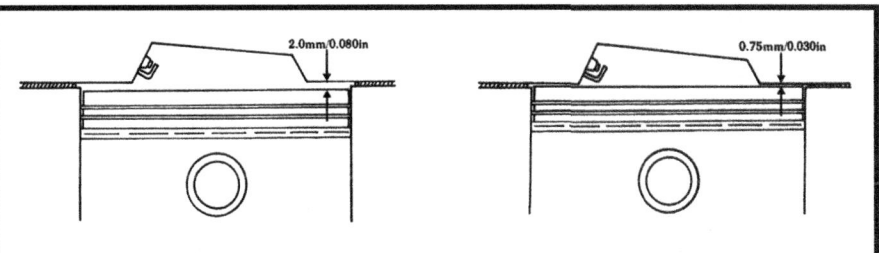

Diagram (left) shows standard gap between piston and cylinder head; diagram (right) shows minimum gap after block planing.

SPEEDPRO SERIES

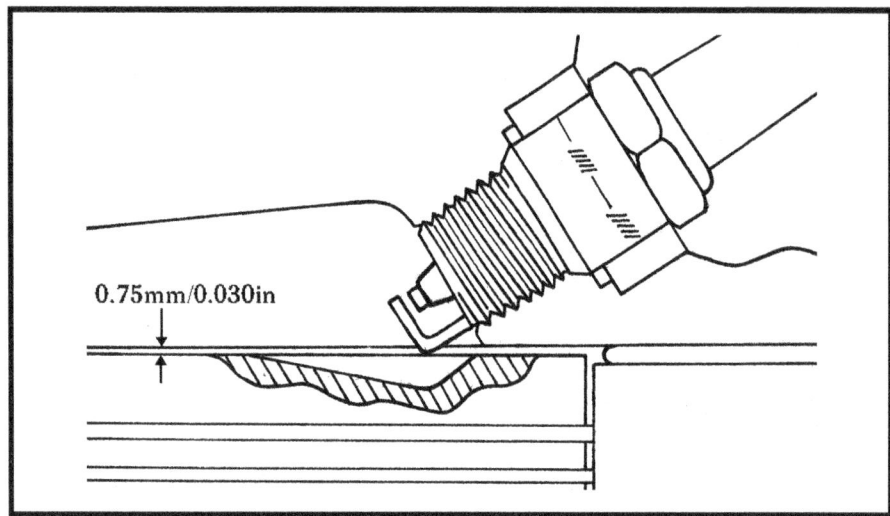

Diagram illustrates the problem of sparkplug electrode/piston contact when a block is planed to give the minimum 0.75-0.9mm/0.030-0.035in clearance between piston and head and the head is planed 2.0mm/0.078in. In this situation a recess 3mm/0.120in deep must be made in the piston crown as shown.

0.005in/0.125mm is the normal amount required to clean up the cylinder head surface.

The 2000 engine's combustion chamber (which is the largest), for example, has an approximate volume of 48cc. Planing this cylinder head by 1.5mm/0.060in reduces the volume to approximately 41cc. Planing this head 2.0mm/0.080in reduces the chamber volume to 35cc and planing by 2.5mm/0.100in reduces the chamber volume to 32cc.

If a 2000 cylinder head is planed 2.0mm/0.080in and the combustion chamber shape is altered to unshroud the valves (increasing the chamber volume as a direct consequence), it will have a chamber volume of around 40cc (this figure will vary a little depending on the amount of reshaping) but consider 39 to 41 cc to be the likely range.

By planing the 2000 cylinder head by 1.5mm/0.060in but doing nothing else, the compression ratio is increased from the standard 9.2:1 to, approximately, 10.2:1. This is not a big increase but it is still worthwhile.

A considerable amount of material can be removed from the Pinto cylinder head without detriment.

Consider 1.5mm/0.060in to be the *safe maximum* amount of material that can be removed from the cylinder head without there being any possibility of problems.

Caution! The maximum amount to remove from the cylinder head surface, without other related modifications, is 2.0mm/0.080in and this amount will increase the CR of a 2000 engine (without chamber work) to, approximately, 11.3:1. **Caution!** With the cylinder head planed by 2mm the sparkplug earth electrode can be proud of the head gasket deck to the point where the piston crown can contact its earth electrode (will depend on whether or not the block has been planed and the thickness of the head gasket). The solution to this problem when it occurs is to have a spherical recess professionally machined into the piston crown directly under the sparkplug. This will allow clearance for the earth electrode, irrespective of where the electrode is positioned when the sparkplug is tensioned, and will prevent the flame front being too close to the top of the piston.

There is one option which allows the cylinder head to be planed to the *absolute maximum* of 2.5mm/0.100in. You'll have to have specially made sparkplug inserts fitted to the original sparkplug holes; this allows the fitting of smaller (10mm) sparkplugs. Note that with this amount of planing some of the material thickness sections

The spark plug electrode is proud of the gasket face on this cylinder head.

COMPRESSION RATIO

of the head face will be getting a bit thin.

If 2.5mm/0.100in is removed from the cylinder head of a standard 2000 engine, the compression is increased to, approximately, 12:1. The main advantage of planing the cylinder head by this amount is the fact that a high CR is possible while still using flat-topped pistons.

THINNER HEAD GASKET

The standard head gasket is approximately 1.63mm/0.065in thick when compressed. The thickness of various aftermarket gaskets varies, but the thinnest compressed cylinder head gasket is the Felpro blue head gasket which is 1.0mm/0.040in in thickness: a reduction of 0.6mm/0.025in. The increase in compression is significant, particularly in combination with other methods. To ascertain the overall thickness of the compressed head gasket measure the uncompressed gasket with a vernier caliper and deduct 0.1mm/0.004in from this thickness.

Reducing the thickness of the cylinder head gasket on a 2000 engine by 0.125mm/0.010in reduces the amount of combustion space by 1.63cc. So using a Felpro gasket reduces the combustion chamber's volume by 4cc, increasing the CR of a 2000 engine from 9.2:1 to, approximately, 9.7:1. Note that the Felpro gasket is made for the 2000 engine and has an appropriate cylinder bore aperture of 93mm: the gasket can be used on the smaller Pinto engines, but the potential compression gains will be offset by the slight compression loss caused by the gaskets 'oversized' cylinder aperture.

A considerable amount of material can be removed from the Pinto cylinder head gasket face without detriment.

BLOCK PLANING

Major gains are made if the block is planed. The reason for this is that the whole cross-sectional area of the bore is removed for every increment of planing, but when the cylinder head is planed by the same amount, only the smaller cross-sectional area of the combustion chamber shape is removed. Block planing is almost always carried out on a bare block when the engine is being rebuilt.

Note that the top of the piston on any standard engine is characteristically 0.5mm/0.020in down from the top of the block. The *minimum* clearance allowable between the top of the piston and the head surface of the combustion chamber is 0.75-0.9mm/0.030-0.035in.

Caution! There is a proviso on this size which is that this amount of piston crown to cylinder head clearance at top dead centre (TDC) is designed for an engine turning a maximum of 7500rpm. Another part of this proviso is that the engine must not have excessive piston to bore clearance. The higher the rpm the engine is turned, and the greater the piston to bore clearance, the more piston crown to cylinder head clearance is required. This is because the higher the rpm the more the connecting rods stretch via the forces acting on them. The greater the piston to bore clearance, the more the piston 'rocks' in the bore and the piston crown is effectively higher in this 'rocking' situation (canted over at TDC). Allow a further 0.125mm/0.005in on 0.005in to 0.006in piston to bore clearance engines, and a further 0.125mm/0.005in on engines turning up to 8000rpm. This means that the pistons can only be brought up flush with the top of the block on some engines, otherwise the piston crowns might well hit the cylinder head!

If a Felpro gasket is going to be fitted the block can be planed to a maximum of 0.75mm/0.030in which will see the tops of the pistons protruding 0.010in/0.25mm above the top of the block deck. Removal of this amount of material from a 2000 engine block's deck will reduce the combustion chamber volume by, approximately, 5cc.

Planing a 2000 block 0.75mm/0.030in, with no other changes, increases CR from 9.2:1 to 9.9:1. If a Felpro cylinder head gasket is fitted to this engine too, the CR will be increased from 9.2:1 to 10.5:1, which is generally ideal for use with 97 octane fuel.

If a head gasket thicker than the Felpro item is going to be used, the amount of material that can come off the block can be increased proportionately. **Caution!** This means that any subsequent head gasket that is fitted will have to be at least the same thickness, otherwise the top of the piston will come too close to the flat surface of the cylinder head.

Consider having the tops of the pistons 0.25mm/0.010in above the block as a maximum so that any head gasket can be fitted in the future.

You can assume that for every 0.25mm/0.010in removed from a 2000 engine block, the combustion area volume is reduced by 1.63cc.

The bore tops *must* be chamfered after block planing so that pistons and rings are easy to fit without damage, and all holes, especially the head bolt holes, must be deburred by countersinking after grinding.

RAISED TOP PISTONS

There are plenty of raised top forged and cast raised top pistons available which will allow very high CRs. Raised top pistons fill the combustion

SPEEDPRO SERIES

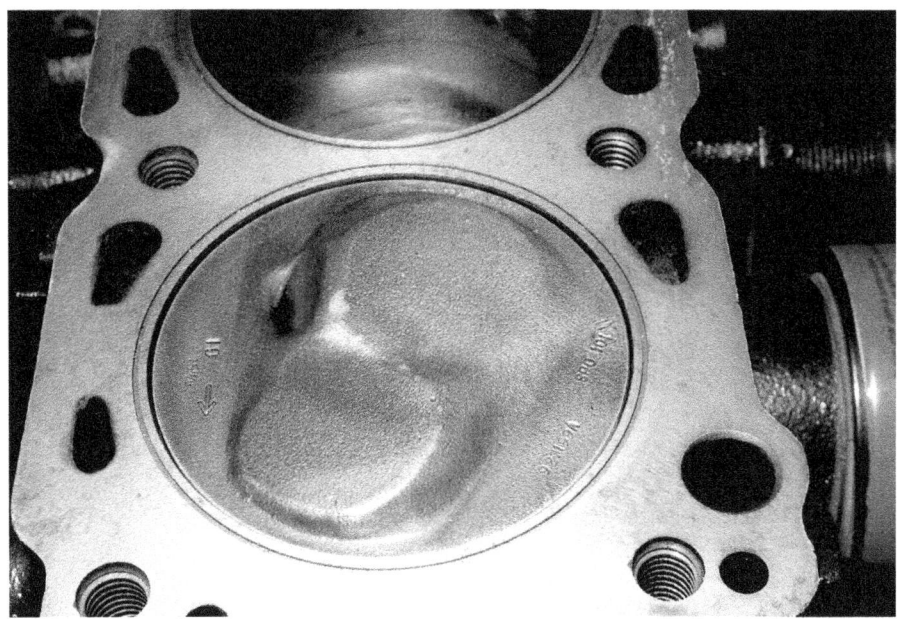

This raised top piston's contour is quite high and is just acceptable for use in Pinto engines. Piston crown has been radiused off by hand filing adjacent to the spark plug recess. Note also that this engine's block has been 'O' ringed to improve the reliability of the cylinder head gasket.

chamber space by varying amounts (heightwise) but if the contouring of the piston crown is too high, flame propagation is restricted. As a consequence the use of very high contour raised top pistons is not recommended for Pinto engines. The majority of high-performance Pinto engines use a combination of block planing and/or head planing and head gasket thickness to achieve the required compression ratio.

'O' RINGING BLOCKS

While head gasket failure is not usually a problem on Pinto engines, it can become one when the compression ratios are over 11 to 1. In the interests of maximum cylinder head gasket reliability, many blocks are grooved, using a milling machine or a hand operated tool designed to do this work, to take high tensile wire (piano wire). The end of the wire that gets

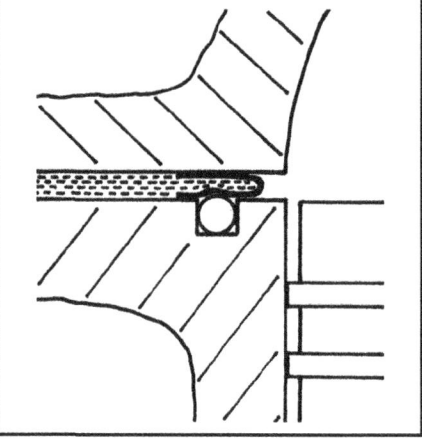

The 0.036 inch diameter wire fitted to the groove, is 0.005inch proud of the block.

fitted into the groove is hand filed square first and the end of the wire that butts up to it is hand filed square to suit, so that there is effectively no appreciable gap between the ends of the wire when the wire is fitted into the groove. The wire protrudes above the block deck surface 0.005 inch/0.125mm for rigid gaskets like Felpro ones, and 0.007in/0.175mm for standard composite gaskets.

COMPRESSION RATIO SUMMARY

The majority of high-performance Pinto engines can have between 10 to 1 and 10.3 to 1 compression for use with 95 octane fuel or from 10.3 to 1 to 10.8 to 1 using 97 octane fuel. Higher compressions than recommended here can be used provided there is no 'pinking'/pinging with 36 or 38 degrees of total spark advance. This is the criteria for determining the amount of compression you can have in your engine. The following list details the various methods of raising CR and records the CR (approximately) given by each. Note that the examples listed pertain to the 2000 engine but the 1600 and 1800 engines will have similar compression ratios after receiving the same treatment.

- Cylinder head planing (no other changes) -
- 1.5mm/0.060in = 10.2:1.
- 2.0mm/0.060in.= 11.3:1.
- 2.5mm/0.100in = 12:1.
- 1.0mm/0.040in head gasket (no other changes) = 9.6:1.
- 1.0mm/0.040in head gasket + planed head 1.5mm/0.060in = 10.7:1.
- 1.0mm/0.040in head gasket + planed head 2.0mm/0.080in = 12.0:1.
- 1.0mm/0.040in head gasket + planed head 2.5mm/0.100in = 12.8:1.
- Large (Group 1) valves + unshrouding (approx. 40cc chamber capacity) + head planed 2.0mm + standard head gasket = 10.4:1.
- Large (Group 1) valves + unshrouding (approx. 40cc chamber capacity) + head planed 2.0mm + 1.0mm/0.040in head gasket = 10.9:1.

COMPRESSION RATIO

- Tops of pistons 0.25mm/0.010in above block + standard head gasket (1.55mm/0.065in) + standard unplaned cylinder head = 9.5:1.
- Tops of pistons 0.9mm/0.035in above block + standard head gasket (1.55mm/0.065in) + standard unplaned cylinder head = 10.0:1.

Caution! If a block is to be planed, irrespective of how much it will be planed and the thickness of the compressed head gasket there *must* be a minimum clearance at TDC of 0.75-0.9mm/0.030-0.035in between the piston and head surface. A further consideration is that, if a standard gasket is used and the block planed to suit, only that thickness of head gasket can be used now and in the future.

Here's a range of modifications and resulting (approximate) CRs -
- Tops of pistons 0.25mm/0.010in above block + 1.0mm/0.040in head gasket + standard unplaned cylinder head = 10.0:1.
- Tops of pistons 0.25mm/0.010in above block + 1.0mm/0.040in head gasket + cylinder head planed 1.5mm/0.060in = 11.3:1.
- Tops of pistons 0.25mm/0.010in above the top of the block 1.0mm/0.040in head gasket + cylinder head planed 2.0mm/0.080in = 12.8:1.
- Tops of pistons 0.25mm/0.010in above the top of the block + 1.0mm/0.040in head gasket + cylinder head planed 2.5mm/0.100in = 13.7:1.
- Tops of pistons 0.25mm/0.010in above the top of the block +1.0mm/0.040in head gasket + cylinder head planed 2.0mm/0.080in + large valves + unshrouding (approx. 40cc chamber capacity) = 11.6:1.

The recommendation is to plane the cylinder block to increase CR. The block is planed by the amount necessary to allow 0.75-0.9mm/0.030-0.035in clearance between the piston tops and the cylinder head surface (with your choice of gasket at compressed thickness). Then, and only then, have the head planed, if necessary, to achieve the desired CR.

Simply planing the cylinder head to increase the compression is not a very satisfactory method overall, although it is an expedient one.

Chapter 7
Camshafts

STANDARD CAMSHAFT
Standard Pinto camshafts are all quite similar with durations ranging from 256 to 268 degrees, and with slightly different valve opening and closing times. Standard camshafts are all very mild in terms of valve action but are very suitable for general purpose, normal use. The valve lifts provided by the standard camshafts are between 9.5mm/0.375in and 10.0mm/0.390in (approximately) – low compared to most high-performance grinds. The standard camshaft has a core diameter size of 27.5mm/1.080in, a camshaft lobe base circle diameter of 30.4mm/1.196in and a heel to toe measurement of around 36.4mm/1.431in.

The standard camshaft is quite suitable for milder high-performance road-going applications where the engine requires good low rpm torque and good power production up to 6300rpm. The suitability of the standard camshaft for milder high-performance applications is generally underrated, so changing to an alternative high-performance profile should not be a step taken lightly. The reliability of the standard camshaft and rockers, coupled with reasonable valve spring pressure (145 pounds 'over the nose'), is about as good as it gets.

To improve the mid-range and top-end performance of a Pinto engine the standard camshaft has to be replaced.

HIGH-PERFORMANCE CAMSHAFTS
There are plenty of different camshaft profiles available for the Pinto engine. The major players being Piper Cams, Kent Cams, Burton Power and Holbay. High-performance camshafts are either reground standard camshafts or made from new blanks. Though the standard camshaft can be reground, there is a limit (based on lobe size) to what can, successfully, be 'put on' to it so, beyond that limit, new blanks are used. There is also a possibility that the standard cam lobe will not be hard enough after being reground.

The actual profile of a high-performance camshaft differs considerably from that of a standard camshaft over and above the duration and lift. The rate of valve opening and closing is as fast as possible (based

Standard camshaft lobes and the core diameter (arrowed) of 27.5mm or 1.080in.

CAMSHAFTS

Standard camshaft on the right and a performance camshaft on the left. Core diameter is smaller on the performance camshaft and so is the base circle. The toe of the camshaft lobe on each camshaft is in the same place (approximately) in relation to the camshaft axis.

on the rocker and valve spring action being able to follow the camshaft lobe) and the valve is kept in the area of full lift as long as possible.

Camshafts manufactured by high-performance camshaft manufacturers will have a core diameter in the vicinity of 26.5mm/1.040in. This means that when a high-performance camshaft lobe is ground, the core diameter is smaller (not by much) than the final lobe base circle diameter, making it easier to grind the lobe as the core never has to be undercut.

When the core of the standard camshaft is undercut there is the possibility of a sharp corner – which can lead to fracture (cracking) – being formed. To remove this corner the camshaft core is usually ground down so that the core is smaller in diameter than the base circle diameter. The standard camshaft's core diameter can be ground/turned down to 25.4mm/1.000in and this allows just about any profile to be added but entails a lot of extra work so it's frequently less expensive to use a new blank.

It's desirable for the camshaft core to be as large in diameter as practicable so that the camshaft is as rigid as possible. Good blanks are not usually less than 26.5mm/1.040in in diameter.

Camshafts made from new blanks are superior to reground standard camshafts, the main advantage being that there is more material on the blank's lobes to start with, and the lobes can be maximum sized. The base circle diameter is also larger than it would be if a standard camshaft was reground. In addition, the hardness of the lobes on new blanks is always very closely monitored to ensure freedom from lobe failure.

The following list gives some useful camshaft dimensions. The dimensions for standard camshafts are accurate, while those for new blanks and reprofiled standard camshafts are all approximate. Heel to toe dimensions do vary because of differences in camshaft lift and, as a consequence, the figures given are loose approximations.

- Core diameter -
Standard - 27.5mm/1.083in.
Blank - 26.5mm/1.040in.
Reprofiled std. - 25.5mm/1.003in.

- Base circle diameter -
Standard - 30.4mm/1.195in.
Blank - 29.5mm/1.160in.
Reprofiled std. (largest) - 28.5mm/1.125in.
Reprofiled std. (smallest) - 26.8mm/1.052in.

- Heel to toe dimension -
Standard - 36.4mm/1.431in.
Blank - 36.9mm/1.452in.
Reprofiled std. (largest) - 34.8mm/1.370in.
Reprofiled std. (smallest) - 34.4mm/1.352in.

CHOOSING A CAMSHAFT

The range of camshafts available is categorized by duration and, to a

SPEEDPRO SERIES

lesser extent, by valve lift. The overall situation is that the more duration a camshaft has, usually, the more lift it will have.

There is frequently considerable confusion over what camshaft to use in a high-performance engine. The tendency is to over-cam (too much duration) and end up with an engine that has insufficient low rpm torque. Pinto engines, however, are less sensitive to long duration than other types and, in 2000 engines, quite 'wild' camshafts (310 to 315 degrees) can be made to pull very well from as low as 2500rpm. These camshafts do, however, give a very rough idle with the idle speed usually needing to be set between 1200 and 1500rpm.

Be realistic in terms of what rpm range is going to be used. Nothing is lost at the bottom end of the rpm range by having a so-called 'mild' camshaft. Conversely, high rpm (over 6000rpm) performance is lost by having a camshaft that does not have enough duration. There is *no* advantage in having more duration than is absolutely necessary. There is little point in fitting a camshaft that has a rev range too much beyond the capability of the particular engine (engine not mechanically strong enough to use those revs). Further to this, the Pinto cylinder head configuration (the inlet port) prevents this engine from continuing to produce 'urgent' power above 7300 to 7500rpm (volumetric efficiency peaks at this engine speed). A slightly milder camshaft (310 degree) can often prove to be far more responsive at low rpm (3500 to 6000rpm) without loss at the top of the range (7500rpm) and can, for all practical intents and purposes, offer a distinct advantage over a longer duration (320 to 330 degree) camshaft.

Listed in this book are the common profiles as made by the prominent camshaft grindings in England. Note that there are many very good camshaft manufacturers and camshaft re-grinding firms around the world who either make or re-grind camshafts. Just because a camshaft re-grinding business is a one man band operation does not mean that his product is not as good as one of the larger manufacturers. Discuss your requirements with your nearest supplier of camshafts, or local camshaft re-grinding concern, as they may well have a very suitable profile for your particular application.

The following list gives a range of camshaft durations, their capabilities and amounts of lift suitable for naturally aspirated engines. In the interests of valve train reliability and being able to use near 'drop in fit' valve springs, avoid valve lifts of over 12.5mm/0.495in.

Camshafts for ALL of these have parameters which for all round general camshaft selection should be adhered to so that you get the type camshaft that you really do need in your particular application. It's a proven fact that road going engines should not have a camshaft fitted to them that opens the exhaust valve earlier than 70 degrees before bottom dead centre (BBDC), while the 'wildest' of racing camshafts should not have the exhaust valve opening earlier than 85 degrees before bottom dead centre (BBDC) under any circumstances as it is just not necessary for up to 7500rpm use. This, therefore, is the range of exhaust opening points to consider from 'mild to wild'. The idea of keeping the combustion pressure working within the cylinder for as long as possible, within reason, is a well founded principle.

The inlet valve should not close any later than 85 degrees after bottom dead centre (ABDC). Actually, no later than 80 degrees is more ideal, though this is not always possible with some camshafts. Certainly 85 degrees is the limit. The milder action camshafts can close the inlet valve as early as 65 degrees after bottom dead centre (ABDC).

The 'overlap' (that's the number of degrees where the exhaust valve and the inlet valve are open at the same time) needs to be kept to 60 degrees for 'mildest' camshafts and 95 degrees for the 'wildest'.

There are several camshafts from both Piper Cams, Kent Cams, Burton Power and Holbay that fall within these boundaries and some are named in the text that follows.

• 260/275 degrees of duration - 10.0-12.0mm/0.390-0.470in lift.

These camshafts have a smooth idle, excellent low end performance and maximum power being produced at, approximately, 6300/6500rpm. Valve spring pressures can be up to 150 pounds 'over the nose' with 140 pounds being quite acceptable.

The camshafts from Piper Cams, for example, that fall into this category are, in order of 'mild to wild', the HR270/2R and the 4PHY3K. The camshafts from the other companies are: Kent Cams FR31, Holbay 4011 and Burton Power BF134.

• 280/305 degrees of duration - 11.0-12.5mm/0.430-0.490in lift.

These are middle of the range camshafts which give rough idles (some worse than others) and produce power from 2000rpm to 7000rpm. These camshafts produce very good mid-range power with good torque and are within the range of durations

CAMSHAFTS

that the majority of high-performance Pinto camshafts will be selected from.

Using valve springs with a minimum of tension is a sound idea as component wear (rocker pads and camshaft lobes) will be minimized. Consider 145/165 pounds of 'over the nose' valve spring pressure to be optimum. Obviously, enough pressure must be applied to avoid 'valve bounce.' Camshafts with lift of more than 11.0mm/0.433in will require single valve springs (offering the same 'over the nose' pressure) which allow more valve spring compression before coil bind; alternatively, the standard valve retainer will have to be turned down by 1.2mm/0.047in to give more coil clearance.

For road-going high-performance engines for which absolute reliability is essential, avoid camshafts with lifts over 11.0mm/0.43in and 'over the nose' valve spring pressures in excess of 160 pounds. Ensure that the rocker geometry is as prescribed in the relevant chapter and limit the engine's maximum rpm by fitting a rev limiter.

The camshafts made by Kent Cams, for example, from 'mild to wild' in this category are the FR32 and the RL31. From Holbay the 401B, and from Burton Power the BF30.

There is another category of camshaft which should not be dismissed. These camshafts are combination grinds which have long inlet durations and shorter exhaust durations. The object of the exercise is to have as much inlet duration as posible (good cylinder filling/volumetric efficiency) and just enough exhaust duration but have as late an exhaust opening point as possible (cylinder pressure action on the top of the piston for as long as possible). The Kent Cams RL30 does this, as does the Holbay 4011C (very good camshafts!).

• 305 to 325 degrees of duration - 12.0-12.5mm/0.470-0.495in of lift (12.5mm is the maximum, but note that few camshafts truly make the advertised lift).

This is the top end of the duration range and there are several well-known camshafts that fall into this category and all of them have very rough idles. There is not much point in fitting a camshaft with a longer duration camshaft than 325 degrees. These camshafts can be quite deceptive in the level of low rpm high-performance they can deliver. Good power delivery can be available from as low as 2500rpm through to 7500rpm. Consider these to be all out competition camshafts.

Examples of this specification of camshaft is the Group 1 profile as made and called the GP1 by Kent Cams and the GP1K or the slightly different phasing of the GP1sK by Piper Cams. The Group 1, 47-85-85-47 phased camshaft is an excellent all-round camshaft.

Examples of this camshaft specification range are Kent Cams RL32 and GTS3 and Piper Cams HR320K and 4PHR3K. Holbay make their 4011F grinds (two) and Burton Power their BF63. The Group 1 camshaft is an excellent all-round camshaft of the full race variety. The timing events are near the limit and, while the lift is perhaps slightly less than some others, it's a reliable camshaft because its full lift valve spring pressure can be less. This camshaft is best used with single valve springs that have an 'over the nose' pressure of 180 pounds (maximum revs = 8000rpm).

Most long duration camshafts have lifts of up to 12.5mm/0.495in and will require dual valve springs to avoid coil bind, but these may exert more poundage than is strictly necessary. **Caution!** Be aware that camshaft and rocker pad wear increases with the use of higher valve spring pressures. Consider 180-185 pounds 'over the nose' to be the *maximum* if relatively good reliability is required.

Because of the usual strength limitation of a standard connecting rod/cast piston equipped engine, consider the Kent Cams RL30 and RL31 to be an ideal type of camshaft to fit to a Pinto engine in view of the maximum reliable rpm available – especially in terms of obtaining the best possible pulling power from low down in the rev range combined with a very broad power band through to 7200 to 7500 rpm.

The Kent Cams RL30 camshaft has timing events that see the inlet valves opening 47 degrees BTDC and closing 77 degrees ABDC; the exhaust valves opening 71 degrees BBDC and closing 41 degrees ATDC. The valve lifts of this camshaft are 12.82mm/0.505in for the inlets and 12.34mm/0.486in for the exhaust. The durations are 304 degrees for the inlet valves and 292 degrees for the exhausts.

The cam's phasing means that the exhaust valve is opened as late as possible so that the cylinder pressure is acting on the piston crown for as long as possible. This feature means that the power stroke is continued for longer than with other camshafts: the trade-off is that the cylinder will not have had time to clear before the piston starts rising, but this only applies at high rpm (starts above 7300rpm). If large (38mm) exhaust valves are fitted to the cylinder head and the exhaust ports enlarged (maximum ported) to suit, this problem does not really affect the engine performance until 7500rpm (use

SPEEDPRO SERIES

41.2mm/1.625in or 47.6mm/1.875in diameter primary exhaust piping).

This cam closes the inlet valve as early as possible in the interest of maximum cylinder filling. This results in a camshaft which has about 90 degrees of overlap making the engine 'lumpy' at idle.

The idea behind this cam design is good and it certainly works, but don't expect an engine fitted with this type of camshaft to have good performance over 7500rpm compared to the performance available between 3000 and 7500rpm.

COMPETITION ENGINES

For competition engines that are race only, the high spring pressure situation is not such a big problem because the engine is not used so much. Also, in the context of competition, the replacement of the camshaft and rockers at relatively frequent intervals is accepted as fair wear and tear. Irrespective of what camshaft is used and what rpm the engine is turned to, use the least full lift valve spring pressure possible to avoid excessive lobe and rocker wear.

For out-and-out competition engines there are camshafts available with up to 330 degrees of duration and valve lifts of 12.6mm/0.500in to 13.25mm/0.520in but, take note, in practice, these radical cams are seldom much more effective overall than with the Group 1 camshaft or equivalents. With very high lifts, valve spring bases may have to be machined so that there is sufficient clearance between the coils at full lift.

The effective maximum power rev limit for the larger Pinto engines is 7800rpm. With a modified standard type of cylinder head, featuring large valves and slightly enlarged ports, it's usual to have good power production to about 7800rpm (seldom more, irrespective of the camshaft used!).

Race applications will frequently see an engine required to rev well beyond the power band without (via a limiting device) cutting out. In such cases, the valve spring pressure will have to be raised to suit anticipated revs but expect higher camshaft lobe and rocker wear as a consequence. Applications like this should use rockers of the segmented type (hard pads) fitted to resist wear, perfectly set up geometry and valve springs giving 'over the nose' pressures of 200 pounds.

Irrespective of what the engine is being used for, all engines should have some form of rev limiting. This can be by way of a 'governor rotor' as supplied by Bosch for their distributors, or an alternative electronic limiting device or, better still, both! The Bosch rotor device is inexpensive.

Camshaft data requirements

Irrespective of what camshaft is to be used, there is a certain amount of cam-related information that is needed so that the engine can be timed correctly. Ultimately, it's only essential to know the rocker to camshaft clearance (tappet clearance) and the cam's position (in degrees) at full lift of the inlet valve of number one cylinder though other information can be useful.

As an example of the sort of information you can expect to receive with your new/reprofiled camshaft, here's the information that comes with the Group 1 specification camshaft - Timing events are -

Inlet opens 47 degrees BTDC and closes 85 degrees ABDC.

Exhaust opens 85 degrees BBDC and closes 47 degrees ATDC

This gives 312 degrees of inlet duration and 312 degrees of exhaust

Adjustable ('vernier') camshaft sprocket/wheel.

duration. The inlet timing is at full lift at 109 degrees ATDC and this is the recommended position, but try the camshaft at 110, 111 and then 112 degrees because this camshaft always seems to make the engine go better timed at 112 degrees.

Inlet valve lift is 12.26mm/0.483in and exhaust valve lift is 11.98mm/0.472in. The tappet clearance is 0.20mm/0.008in for the inlets and 0.25mm/0.010in for the exhausts.

This is usually the minimum amount of information supplied and it lacks the recommended valve spring pressure. Ask for the recommended seated and 'over the nose' valve spring pressure to attain the rpm you require, including a safety and over-speed margin.

Holbay roller camshafts

Holbay make four roller camshaft kits for Pinto engines under coding 509T, 509M, 509R and PR62 with two of them being of special interest (that's the 509M and the 509R). This is pretty specialised gear and works on a slightly different principle to the standard rocker and camshaft lobe setup. The rockers, for a start, are based on a 2 to

CAMSHAFTS

Holbay's roller camshaft lobe.

1 ratio. The roller is halfway between the fulcrum point and the roller tip in contact with the top of the valve. The distance from the fulcrum to the middle of the roller is 31.0mm/1.220inch, compared to the approximate 37.0mm/1.455inch of the standard rocker. The roller is positioned directly under the vertical centre line of the camshaft and more or less maintains this position through to maximum lift (it moves closer to the rocker fulcrum by about 0.75mm/0.030inch actually). The roller tip follows the centre line of the valve stem down exactly. The fulcrum is positioned where it is in order to do this.

While the lobe of the Holbay camshaft may look a bit low on lift it doesn't work with the same ratio (looks can be deceiving). The current range of Holbay camshafts all work with the standard late camshaft bearings or the needle roller assembly of days gone by. Holbay recommend using the needle roller main bearings. The journal sizes of the two types of camshaft are not the same.

The 509M camshaft is listed as being a rally camshaft and has 300 degrees of duration and 0.415inches/10.5mm of valve lift. The inlet valve opens 42.5 degrees before top dead centre (BTDC) and closes 77.5 degrees after bottom dead centre (ABDC). The exhaust valve opens 77.5 degrees before bottom dead centre (BBDC) and closes 42.5 degrees after top dead centre (ATDC).

The 509R camshaft is listed as being a race only camshaft and has 308 inlet/312 exhaust degrees of duration and 0.437inch/11.0mm of valve lift. The inlet valve opens 56 degrees before top dead centre (BTDC) and closes 76 degrees after bottom dead centre (ABDC). The exhaust valve opens 74 degrees before bottom dead centre (BBDC) and closes 54 degrees after top dead centre (ATDC).

Fitting a camshaft is quite straightforward, in that it's as per standard if it is a non-needle roller main bearing camshaft, but the fitting of the rockers is not (Holbay offer two types of camshaft, one suits the standard camshaft bearings and the other is a needle roller conversion). Each rocker has to be lined up with the camshaft lobe it is going to run against. Extreme care is necessary to get this right as well as getting the securing bolt (one per rocker assembly) fully tensioned.

The rollers of each rocker have to be aligned using bearing blue. This means fitting each rocker assembly individually onto the cylinder head in a specific order and checking the alignment of each rocker's roller by smearing bearing blue over each roller and turning the camshaft through one revolution and checking to see whether the bearing blue has transferred 100% across the surface of the camshaft lobe. This is 'fiddly' work and has to be got exactly right. The securing bolts are torqued up to 45 foot pounds. Note that the standard thread size is larger than the Holbay supplied bolt. Original pivot holes in the cylinder head all need to be heli-coiled 1/2 inch UNF.

The base circle diameter of the camshaft lobes and the rocker assembly as a whole are sized so that with the rocker bolted onto the cylinder head everything is essentially in the right position for operation.

The valve clearance is set by undoing the securing bolt and nut which holds the rocker's valve stem roller. This fixed for operation roller, is on an eccentric and has a total amount of clearance adjustment of 1.0mm/0.040inch. If you run out of adjustment at the valve stem roller, the rocker 'yoke' will have to be packed up to bring the rocker tip into the adjustment range. It is all quite close, and packing up the rocker yoke does not result in incorrect geometry.

The Holbay valve spring retainers must be used, since they are flatter than the standard ones. The underside of the rocker has been milled out to clear the retainer as it is.

The Holbay rocker assembly is a complicated system but it's not that complicated. Once set up correctly the system is reliable. It's drawback is the initial cost compared to the conventional Pinto camshaft and rocker system componentry.

Current rocker assembly is the third generation type. The camshaft can be turned by hand when fully assembled via the resulting low friction aspect of the roller system.

CAMSHAFT TIMING

The standard camshaft timing drive pulley is fixed in one position relative to the cam's centre axis by a key and keyway system. For high-performance engines an adjustable camshaft pulley is *essential*. When the block and/or head has been planed, the camshaft timing will automatically be retarded so the fitting of an adjustable pulley can correct this and allow infinite adjustment of cam timing.

The camshaft timing is usually set up using a degree wheel. Alternatively, the crankshaft pulley/damper is marked precisely with the full lift timing point.

If a timing disc is to be used it needs to be mounted on the crankshaft and then set correctly in relation to TDC of number 1 piston. Note that the full lift timing position always relates to the inlet valve of number 1 cylinder.

A dial indicator is used to measure the exact point that the valve reaches full lift. With the Pinto's cast iron head the dial indicator is usually fitted to a magnetic standard so that the dial indicator is positioned directly above the valve spring retainer.

The first reading is taken by rotating the engine clockwise until the valve just reaches full lift (dial indicator needle stops moving). A degree reading is then taken off the timing disc and recorded. The crank is then rotated a little further clockwise past the full lift position before being turned anti-clockwise until the dial again indicates the inlet valve has reached full lift. Another reading is taken from the degree disc. The true full lift position is exactly between the two readings taken from the timing disc.

If an adjustment of cam timing is necessary slacken the vernier cam sprocket bolts, set the camshaft so that the inlet valve is at full lift and then rotate the crankshaft clockwise until the correct number of degrees is read on the timing disc before tightening the vernier sprocket bolts. Check the new setting as previously described.

The initial camshaft setting should be to it's manufacturers recommendation. After testing the engine, reset the camshaft to 3 crankshaft degrees in advance and 3 crankshaft degrees retarded from the recommended setting and test the engine at both settings. Use the setting which works best for your application. Note that there is rarely any advantage in going beyond 3 degrees advanced/retarded.

The recommended degree setting, plus 3 degrees of advance and retard, can be permanently marked on the crankshaft pulley – this will make it easy to check and reset camshaft timing when the engine is installed in the car.

CAMSHAFT SUMMARY

The following applies to naturally-aspirated engines fitted with a twin barrel downdraught or sidedraught carburettor, or a pair of twin choke sidedraught Weber or Dellorto carburettors (one choke per cylinder).

Standard camshaft have lifts between 9.5mm/0.375in and 10.0mm/0.393in.

Performance camshafts have lifts between 10.0mm/0.393in and 13.5mm/0.530in.

For maximum reliability choose a camshaft with lift up to 11.0mm/0.433in.

For maximum rpm requirements of 6300rpm use the standard camshaft.

For up to 6300rpm use 'over the nose' valve spring pressures of 145 pounds.

Use durations of 275 degrees for applications up to 6700rpm.

Use durations of 305 degrees for applications up to 7200rpm.

Use durations of 315 degrees for applications up to 7500rpm.

Chapter 8
Valve springs

With Pinto engines it's important to have enough valve spring pressure to ensure that the valves and rockers follow the camshaft profile exactly. More than the required amount of valve spring pressure (with a small safety margin) to prevent valve bounce is not necessary, or desirable, with these engines. The fitting of the softest valve springs possible and limiting the engine's maximum rpm are well founded principles. In the interest of reliability, avoid using 'over the nose' valve spring pressure in excess of 180-185 pounds with standard type rockers (depending on the camshaft, 8000rpm is often possible with this pressure). Check that the rocker geometry is correct (see relevant chapter) when the rocker is in the full lift position.

There's a huge array of valve springs available from many companies. Irrespective of who makes the spring, the correct choice can be made if the exact requirements for your application are correctly established. The valve springs that can be used on Pinto engines can be categorised into the six following types –

1 - Low tension-rated standard single valve springs

2 - Medium tension-rated standard single valve springs

The first two valve springs are Ford made. Both valve springs fit into the standard 36.0mm/1.417in fitted spring height of the standard head. Both valve springs coil bind at the same height and, when used with standard camshafts, are compressed by a maximum of 10.0mm/0.393in. The first valve spring gives approximately 135 pounds of 'over the nose' pressure, while the second spring gives approximately 165 pounds of 'over the nose' pressure.

3 - Medium tension-rated dual valve springs

The third valve spring type is the standard-rated dual replacement valve spring (FAI, for example) that offers approximately 135 pounds 'over the nose' pressure with 10.0mm/0.393in valve lift and approximately 157 pounds 'over the nose' at 12.0mm/0.472in lift. These springs allow a maximum valve lift of 12.7mm/0.5in and with, approximately, 165 pounds 'over the nose.' These valve springs only offer more spring tension than the standard springs because they have a lower coil bind height. By way of explanation, you need to understand that all springs have a changing rate of resistance as they are compressed – the greater the compression, the greater the resistance. These dual springs allow more camshaft lift without increasing the 'over the nose' poundage. Standard springs have 165 pounds at 26mm/1.023in compared to the dual's 165 pounds at 23.3mm/0.917in.

4 - Low tension-rated single valve spring (allows high lift)

These valve springs are a 'drop in' fit and give a seated pressure of 50 pounds at the standard 36.0mm/1.417in installed

SPEEDPRO SERIES

Left to right - standard valve spring, low tension valve spring (which allows more lift), high tension single valve spring and high tension dual valve springs.

spring height. They coil bind at 23.0mm/0.905in but will compress usefully down to 24.0mm/0.944in when they will give an 'over the nose' pressure of 145 pounds. These valve springs are ideal for high lift (12.0mm/0.472in) camshafts and revs up to 6700-7200rpm (maximum rpm depends on the camshaft profile).

5 - High tension-rated single valve springs

These very strong single springs are not designed to fit into the standard 36.0mm/1.417in fitted height. The reason for using this sort of spring is to get a spring tension with 'over the nose' pressure up around 180-185 pounds. Usually these springs will have a 37.0mm/1.456in fitted height which will mean that the valve retainers will definitely have to be machined down by 1.00-1.25mm/0.40-0.050in to meet this requirement. The spring bases in the cylinder head can also be machined, but this involves extra work. These springs are only for use with camshafts giving up to a maximum of 11.0mm/0.433in lift.

The springs will have a fitted height pressure of 75 to 80 pounds, which is quite high. One thing strong singles like this do not do is to compress usefully closer than 2mm/0.078in off coil bind (they break). Expect the coil bind height to be 23-24mm/0.905-0.944in which means the allowable lift could be either 10mm/0.393in (standard retainer) or 11mm/0.433in (with a machined down retainer).

6 - High tension-rated dual valve springs

Springs such as these are likely to have a free length of 43.0mm/1.692in and a coil bind height of 21.0mm/0.826in. The outside diameter of these springs will be 32.0mm/1.259in and the inside diameter will be 18.0mm/0.708in. With the standard 36.0mm/1.417in installed height the average seated spring pressure will be approximately 73 pounds and at 12.0mm/0.472in valve lift the 'over the nose' pressure will be approximately 210 pounds. If the valve retainer is turned down by 1.25mm/0.050in, the pressure will reduce to approximately 63 pounds and the 'over the nose' pressure at 12.0mm/0.472in will be approximately 202 pounds.

If these particular springs are used with a camshaft which has 13.0mm/0.511in of valve lift, the 'over the nose' spring pressure will be back up to 210 pounds. These valve springs represent the maximum possible pressure and are more than capable of preventing valve bounce to 8500rpm and beyond.

By machining the standard valve retainers (up to 1.25mm/0.049in) and machining the spring seat bases of the cylinder head, valve spring pressures can be reduced to 180-185 pounds which is the recommended maximum for reliability. You'll need to determine at what length the springs give the desired pressure (see section on measuring valve spring poundage) in order to determine how much material to remove from the retainer and cylinder head spring seat. The remachining (spot facing) of the spring base platforms is a common procedure in the modification of Pinto engines which can be carried out by engine machine shops and high-performance engine specialists.

VALVE SPRING DIMENSIONS

The sizes are for springs that will fit directly into the cylinder head and include the extra tolerance that can be gained when up to 1.25mm/0.049in is machined off the underside of the valve retainer.

The outside diameter can be 31.5mm/1.240in, 32.0mm/1.259in, 32.3mm/1.271in or 32.8mm/1.291in.

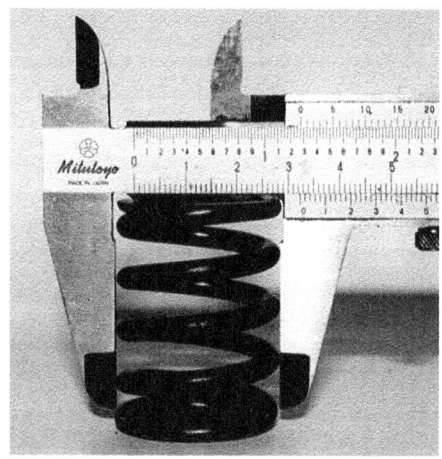

Vernier caliper measuring the diameter of valve spring.

VALVE SPRINGS

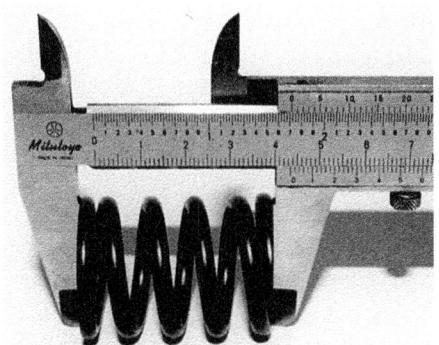

Free-length of valve spring being measured.

Valve spring compressed between vice jaws and vernier caliper being used to measure the distance between the jaws.

The free-length height of the valve springs will usually be 43mm/1.692in, 44mm/1.732in or 45mm/1.771in (perhaps up to 45.5mm/1.791in). This specification can, however, be misleading because free-height is not a guide to what bind height will be; it is, nevertheless, a guide to the suitability of valve springs.

The coil bind height needs to be 21mm/0.826in, 22mm/0.866in, 23mm/0.905in or 24mm/0.944in. High lift camshafts which have 12.5mm/0.492in lift will require valve springs with a coil bind height of 22mm-23mm/0.866-0.0905in or, if the valve retainer is machined, a 24mm/0.944in coil bind height spring could be used. The bind height measurement is found by compressing the valve spring in a vice until it is just clamped solid and then measuring the gap between the vice jaws with a vernier caliper (**Caution!** This is a potentially dangerous procedure (flying spring): have it done by a professional).

Take, for example, some racing dual valve springs. The valve spring is coil bound at 22mm/0.866in and will therefore allow a maximum valve lift of 13.0mm/0.511in. The fitted height is 36mm/1.417in, minus 13mm/0.511in valve lift plus 1mm/0.039in clearance between the coils equals 22mm/0.866in.

If the retainer is machined down 1.25mm/0.050in, the camshaft could have 14.0mm/0.55in of valve lift with the same spring or, if a camshaft was installed which had 12.5mm/0.492in valve lift, there would be 2.0mm/0.078in clearance between the coils instead of 1.0mm/0.039in. If the retainer is machined down the valve spring pressure drops slightly but it will compress a further 1.25mm/0.049in before coil bind.

STANDARD VALVE SPRING DATA

The standard valve springs have a free-length of 44.0mm/1.732in and the fitted spring height in all cylinder heads is 36.0mm/1.418in. The standard valve springs have seated pressures ranging from about 58 to a definite maximum of 68 pounds and, when the valves are at full lift (compressed about 10.0mm/0.390in), they give 'over the nose' pressures ranging from approximately 135 to 165 pounds of pressure.

At full valve lift these standard springs are compressed to a maximum of 26.0mm/1.020in which is well within the range of the springs. Compress these springs down to 24.0mm/0.946in (a further 2.0mm/0.080in) and they become coil bound. Using these valve springs with camshafts that have a true lift of up to 10.5mm/0.412in is possible and acceptable. All valve springs used in Pintos should be tested at the 'over the nose' pressure to simulate when the camshaft is at full lift (see section on measuring spring poundage).

Caution! 10.5mm/0.412in of actual valve lift is the *absolute maximum* that could ever be used with standard springs and, as most high-performance camshafts have more lift than 10.5mm/0.412in, the standard valve springs are not able to be used (too near to coil bind). The poundage that these springs exert when seated or at full lift will allow engine speeds of up to 7300rpm (depending on the lobe profile), but the prospect of running an engine with springs approaching the onset of coil bind is worrying and *must* be avoided.

Caution! Note that some of the following camshafts cannot be

Standard spring retainer on the left and a turned down one on the right. Difference in thickness is 1.25mm/0.050in.

SPEEDPRO SERIES

used with standard Sierra IS valve springs and standard retainers unless the retainers are machined down by 1.25mm/0.050in to increase the spring fitted height to 37.25mm/1.462in.

The camshafts concerned are listed in the Piper catalogue as the BP255K, BP270K, HR255K, HR270K, HR270/2K, HR285K, HR300K, 4PHY1K, 4PHY2K, 4PHY3K, 4PHR1K, WR40K. However, the pressure on the rockers is similar to that of the standard engine so provided the spring retainers *are* machined and rocker geometry is reset to standard specifications, these camshafts will be as reliable as any standard camshaft.

Caution! Any camshaft with a true lift of over 10.5mm/0.414in *must* have the spring retainers turned down.

The following valve spring examples give the specifications of springs within factory ratings -

1 - Sierra IS valve springs
Inside diameter - 32.5mm/1.278in.
Outside diameter - 23.5mm/0.925in.
Free-length - 44mm/1.730in.
Fitted height - 36mm/1.417in.
Seated pressure - 66 pounds (approx.).
10mm/0.393in full lift pressure - 165 pounds (approx.).
10.5mm/0.412in maximum lift pressure - 170 pounds (approx.).
Coil bind height - 24mm/0.945in.

2 - Sierra 1800 valve springs
Inside diameter - 23.5mm/0.925in.
Outside diameter - 31.8mm/1.250in.
Free-length - 47.0mm/1.850in.
Fitted height - 36.0mm/1.417in.
Seated pressure - 63 pounds (approx.).
10.0mm/0.393in full lift pressure - 133 pounds (approx.).
10.5mm/0.414in maximum lift pressure - 137 pounds (approx.).
Coil bind height - 24.0mm/0.945in.

The standard valve lift on the 1800cc Sierra engine is 9.5mm/0.375in and the 'over the nose pressure' is about 130 pounds. All Pinto standard valve springs are able to rev to 6700rpm so, with the standard camshaft having a maximum power range of 6300rpm, the standard valve springs are adequate. The standard camshaft performs much better than it is given credit for.

3 - FAI dual valve springs
(Note, while not a Ford spring, seated and 'over the nose' pressures are within factory ratings and these springs are sold with standard replacement camshaft kits)
Inside diameter - 18.0mm/0.708in.
Outside diameter - 31.0mm/1.220in.
Free-length - 45.0mm/1.770in.
Fitted height - 36.0mm/1.417in.
Seated pressure - 53 pounds (approx.).
10.0mm/0.394in full lift - 135 pounds (approx.).
12.5mm/0.490in maximum lift - 157 pounds (approx.).
Coil bind height - 22.0mm/0.865in.

These replacement dual valve springs give similar pressures to the standard springs but they will compress to 22.0mm, which will allow the use of a high lift camshaft without running excessive spring pressure. Engines fitted with these valve springs and a 12.5mm lift camshaft should have the maximum rpm limited by mechanical or electronic means, or both, to about 7000rpm.

A point of note is the fact that the inner valve spring is shorter than the large main spring. This serves to reduce the overall valve spring pressure at the fitted height as the inner spring is not compressed by much at that point.

Piper Cams, for example, sell several camshafts that will suit these valve springs and the duration range of these 'low lift' camshafts is extensive.

4 - Typical aftermarket low tension single valve springs
Inside diameter - 23.5mm/0.930in.
Outside diameter - 31.5mm/1.240in.
Free-length - 45mm/1.770in.
Fitted height - 36.0mm/1.417in.
11.0mm/0.430in full lift pressure - 130 pounds.
12.0mm/0.470in full lift pressure - 135 pounds.
Coil bind height - 23mm/0.905in.

These valve springs allow more valve lift than standard, but give a full lift 'over the nose' pressure which is the same as the standard 1800 Sierra.

For any alternative valve spring to have the advantage of being a 'drop in fit' and yet offer a gain over the standard valve spring for extra lift, the coil bind height has to be a minimum of 23.0mm/0.905in. This will allow 11.8mm/0.462in of valve lift.

5 - Typical aftermarket high tension single valve springs
Inside diameter - 24mm/0.945in.
Outside diameter - 32.8mm/1.290in.
Free-length - 45.5mm/1.790in.
Fitted height - 37.25mm/1.468in.
Seated pressure - 75 pounds (approx.).
10.0mm full lift pressure - 175 pounds (approx.).
11.0mm full lift pressure - 190 pounds (approx.).
Coil bind - 24mm/0.940in (**Caution!** compress no more than 25.8mm).

The strong single valve spring can be used with medium lift camshafts, but *must* be used with turned down retainers to obtain the maximum permissible lift of 11.3mm/0.443in. These valve springs can be used for competition purposes (if necessary, cylinder head valve spring bases can be machined to reduce spring pressure).

VALVE SPRINGS

6 - Typical aftermarket high tension dual valve springs
Inside diameter - 18.0mm /0.705in.
Outside diameter - 32.0mm/1.255in.
Free-length - 43mm/1.700in.
Fitted height - 36mm/1.417in.
36mm fitted height seat pressure - 75 pounds (approx.).
12.5mm full lift pressure - 205 pounds (approx.).
13.5mm full lift pressure - 210 pounds (approx.).
With turned down retainers -
37.25mm fitted height seat pressure - 65 pounds (approx.).
12.5mm full lift pressure - 189 pounds (approx.).
13.5mm full lift pressure - 203 pounds (approx.).
Coil bind height - 21.0mm/0.825in.

The fitting of dual springs for competition or any continuous high rpm application is *essential*. The coil bind height will normally be approximately 21.0mm-22.0mm/0.830in to 0.864in to allow for a high lift camshaft and the increased tension allow up to 8500rpm. The fitted height of most dual valve springs is more than the standard 36.0mm and may well be listed as 37.0mm/1.455in. The standard retainers can be turned down 1.0mm/0.040in to compensate.

If necessary, the spring bases in the cylinder head could also be machined to gain more clearance to suit valve springs that are longer, but this should not be necessary given the spring choice available.

Holbay make and supply one dual valve spring combination for use with all of its camshafts. These valve springs, called 'Rocket' valve springs, have a free length of 43.2mm/1.700inch and a coil bind height of 19.5mm/0.775inch. These springs are designed to fit with a 1.5mm/0.062inch thick valve spring

Holbay 'Rocket' dual valve spring with Titanium valve spring retainer and Titanium spring base washer. The Titanium valve spring retainer weighs 9 grams.

base washers which locates the base of the valve springs accurately in the cylinder head. Valve spring retainers are available in titanium or steel. These valve springs are NOT designed to be fitted into the standard fitted height size. The cylinder heads have to have the standard spring base spot facings machined deeper, by 2.50mm/0.100inch. This results in a 38.5mm/1.518inch fitted height dimension as opposed to the standard 36.0mm/1.420inch size. The seated pressure is approximately 70 pounds and the 'over the nose' pressure at 10.5mm/0.415inch of valve lift is 180 pounds, at 11.5mm/0.455inch of valve lift it is approximately 190 pounds, whilst at 12.5mm/0.495inch of valve lift it is approximately 200 pounds.

MEASURING VALVE SPRING POUNDAGE

Special, very accurate, machines are available for checking spring pressures but they are very expensive and not all engine reconditioners have them. It is not acceptable to simply buy

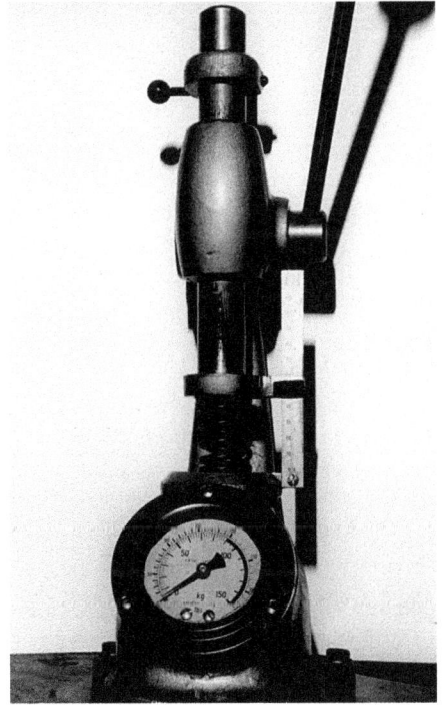

Valve spring pressure testing machine.

valve springs and install them in an engine without knowing approximately what the seated pressure of every valve spring is and, more importantly,

SPEEDPRO SERIES

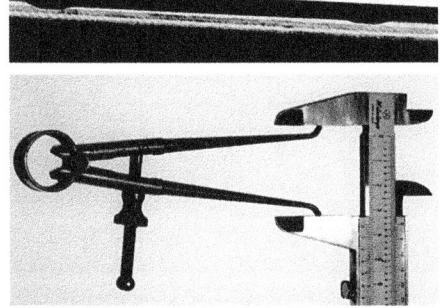

300mm/12in square piece of plywood on the drill press table underneath the bathroom scales. The second piece of plywood is 250mm/10in square and sits on top of the scales to ensure that the load is spread evenly. The spring sits between plywood and drill chuck. The scales must be reset to zero before each measurement. Use dividers to measure spring height and a vernier caliper to take the measurement from the dividers.

Warning! The following procedure is potentially dangerous (flying springs) so have it done by an expert if you are not able to guarantee your own safety. The means of measuring the valve spring pressure described here is not completely accurate but, if care and attention is taken, a sufficiently close approximation is possible.

The equipment required is a drill press, a set of flat bathroom scales, two squares of 12.7mm/0.5in thick plywood (see photos), a vernier caliper, a pair of inside calipers, a pair of dividers and a calculator.

With the spring between the top square of plywood (sitting on the scales) and under the drill chuck, the dividers are set to 36.0mm/1.415in (set the vernier caliper to the size required and use it to set the dividers) and the spring is then compressed to the desired height by winding the drill press handle in the usual way.

With the valve spring compressed to the required height, the dial of the scales can be read (14 pounds per stone/1 pound = 0.4536kg). Good quality scales are normally pretty accurate and by using the same scales for all the measurements, the comparison between the springs will be accurate. Make a note of all heights and pressures for future reference.

what the pressure 'over the nose' is at full valve lift. Excessive valve spring pressure at full lift, especially if coupled with incorrect geometry, will wipe a camshaft lobe off very quickly. **Caution!** You *must* check each spring individually at the true fitted height and the true full lift height.

Caution! For most applications, where reliability is essential, avoid 'over the nose' (full lift valve spring pressures) of more than 170 pounds. This pressure is sufficient to run many profiles up to 8000rpm and 160 pounds is sufficient to run many profiles to 7500rpm. The full lift 'over the nose' spring pressure is more important than the seated valve spring pressure.

It's not possible to give set amounts of valve spring pressure for use with a certain amount of rpm as it is not a pro-rata situation. This is because the camshaft lobe profiles vary considerably, some having easier actions than others, and, as a consequence, will rev higher for less valve spring pressure. You can reliably assume, however, that more angular camshaft lobes need more valve spring pressure than more rounded lobes to achieve the same rpm.

The accompanying photograph shows the radius on the top of the lobe (left-hand side) of the standard camshaft (left) is quite small and the radius (left-hand side) on the lobe of the 285 degree camshaft (centre) is of even smaller radius still, yet this latter camshaft is almost always going to be used with stronger valve springs and more engine rpm than the standard camshaft. The racing camshaft profile (right) has a larger radius (left-hand

Lobe profile comparison - standard camshaft profile (left), 285 degree high lift camshaft (centre) and 312 degree race camshaft (right).

VALVE SPRINGS

Spring fitted height being measured between the valve retainer and spring seat using inside calipers.

side) than either of the others and will definitely be used with more engine rpm than the other two profiles, yet the spring pressure requirement (180-185 pounds over the nose) will not need to be all that high to run at 8000rpm.

ADVERTISED VALVE LIFT

The advertised valve lift is not always realized in practice so a camshaft that is listed as having 11.0mm/0.432in of lift may well be able to be used with standard valve springs as the true lift may be around 10.7mm/0.420in. To check lift, start by measuring the fitted heights of the springs (distance between the underside of the valve spring retainer and the bottom of the spring base's machined recess) to see that all are 36.0mm/1.416in. Check what the valve spring pressures are at 36mm/1.417in compressed height.

When the camshaft is installed, measure the gap, or gaps, between the coils of each valve spring at full lift. Go by the gap available between the centre coil, or coils, at full lift. If a 0.075mm/0.030in gap (or greater) exists there will be no coil binding with this cam/spring combination. The way to find out whether a camshaft which, theoretically, has too much lift can be successfully run in an engine which has standard valve springs, is to assemble it and check it in the prescribed manor. Sufficient clearance between the coils at full lift and the required 'over the nose' spring pressure is what you need.

Further to this, if insufficient clearance exists between the coils at full valve lift the valve spring retainers can be turned down in a lathe by up to 1.25mm/0.050in, which will increase the gap between the coils by the same amount. Note that standard Ford valve spring retainers are hardened, but they can be turned by using a tungsten carbide tool. The fitted height spring poundage will drop by 5 pounds, or so, but if the engine is not going to be revved that hard (over 6300rpm) it's an option which might allow you to use your existing valve springs with such a camshaft.

The minimum 'over the nose' valve spring pressure to ever use is 130 pounds.

Gap between the coils at full valve lift being measured.

Standard valve stem seal (left) and smaller diameter oil seal which will fit inside dual valve springs (right).

SPEEDPRO SERIES

COMPETITION ENGINES

The maximum rpm that a particular set of valve springs will allow also depends, to a certain degree, on the profile of the camshaft lobes. In short, one set of valve springs as fitted to a cylinder head may well allow two different amounts of maximum rpm if two different camshafts are tried in the engine. It is not possible to give a set maximum rpm rating for a valve spring combination, only a close approximation.

If an engine is subjected to continuous high rpm operation, such as in pure racing, valve spring tension is kept on the high side of the known required tension to counteract spring pressure reduction over a period of usage. Camshaft and rocker wear resistance may well have to be sacrificed to a small degree if the highest practicable spring pressure is used. Replacing the camshaft and rockers when a competition engine is totally rebuilt (after a season) is not uncommon.

With a racing engine, there are times when it will be revved well above the effective power band. As an example, this happens when, instead of making a gearchange, the engine is held at full rpm and the revs allowed to rise above the usual change point between two corners when neither gear ratio is quite right. The engine rev-limiting device may well be set to the maximum safe rpm for the engine's internal components, which is higher than the actual maximum power point of the engine. The maximum allowable spring pressure may be used in these cases and high wear rates/reduced reliability accepted as a necessary evil.

On competition engines, valve springs are frequently changed to preclude breakage and also prevent, as far as possible, the inevitable reduction in spring tension brought on by usage which could cause the engine not to rev to the usual maximum. If the original seated pressure and the full open pressures are known, all valve springs can be tested periodically.

Valve springs are made of very good quality material and spring breakage is not common – unless the valve springs are operating in a near coil bind situation. Loss of tension is far more likely than breakage.

When ordering valve springs, find out what the outside and inside diameter of the valve springs is so that you can be sure the springs are compatible with the standard valve retainer. Note that some dual valve springs will not allow the fitting of the large standard valve stem seals (early ones) which are 18.0mm/0.707in in diameter. However, there are many alternative valve stem seals available from other engines which will fit straight on and yet have diameters of 16-17mm/0.629-0.669in (some will need to be trimmed). Consult a parts store or engine machine shop which will be able to supply a direct equivalent. The valve stem size is 8mm/0.314in and the top of the guide is 12mm/0.472in in diameter. The profile of the seal when fitted to the cylinder head must be little or no higher than the standard seal. Consider 200 pounds of 'over the nose' pressure to be the *absolute maximum.*

VALVE SPRING SUMMARY

For all standard camshafts being used up to 6300rpm (and rev-limited), use 135 to 150 pounds of 'over the nose' valve spring pressure.

For high-performance type camshafts (280 to 300 degrees duration) with lifts of 10mm-11.2mm/0.393-0.440in using single valve springs with the rpm limit set to between 7000 and 7500rpm use 'over the nose' valve spring pressures of 165 to 175 pounds.

For high-performance type camshafts (300 to 315 degrees duration) with valve lifts of between 10mm-11.2mm/0.393-0.440in using single valve springs use valve springs with 'over the nose' pressures of 175 to 185 pounds. Maximum revs 8000rpm.

For high-performance type camshafts (280 to 300 degree duration) with lifts from 11.0-12.5mm/0.433-0.492in use dual valve springs with 'over the nose' pressures of between 165 and 185 pounds. Limit revs to 7000-7500rpm.

For competition camshafts (300 plus degrees duration) being used in engines that are being rev limited to 7000-7200rpm with up to 12.5mm/0.492in lift use 'over the nose' valve spring pressures of 140 to 150 pounds.

For competition engines (310 degree plus duration camshafts) with camshaft lifts of 11.0-13.5mm/0.433-0.531in using 7500 to 8000rpm plus, use dual valve springs with 'over the nose' valve spring pressures of not less than 180-185 pounds and avoid poundage's over 200.

Note that some high revving Pinto racing engines (8500rpm plus) are being fitted with camshafts which have over 14.0mm/0.550inch of valve lift and 230 pounds of 'over the nose' valve spring pressure. Engines like this receive frequent maintenance and complete valve train part replacement on a regular basis.

Chapter 9
Rockers & rocker geometry

ROCKER GEOMETRY CRITERIA

The standard rocker geometry is correct for the standard sized camshaft lobe, standard rocker and standard valve stem length. The overall design relies on the rocker pad radius being in a set position at full camshaft lift. This relationship is the common denominator that is often completely lost when the standard camshaft is changed for a high-performance one. The full lift attitude of any high-performance camshaft's rockers should always be the same as the standard camshaft's.

The standard rocker arm.

All camshaft lobes and standard rockers are compatible provided the hardness values of each are to Ford's specifications (some are not). There is a considerable difference in the hardness of rockers, and the only way to test the rocker pad for hardness is on a Rockwell tester: the harder the better, but, for most applications, rockers do not need to be tested, just set up right geometry-wise. **Caution!** Any application that is going to be run with valve spring pressures of 200 pounds 'over the nose' at full lift *must* have the rockers tested in this manner to make sure that all rockers are up to the specified hardness (52 plus Rockwell C).

All camshaft lobes and rockers will eventually wear but, provided the oil is changed regularly and there are no blockages in the oil spray bar, the wear will not be abnormal. Invariably the nose of the camshaft lobe and the rocker pad wear across the rubbing area – predominantly in the centre of the pad, because here the wiping action is slowest and the pressure

Badly worn rocker which had been running with correct geometry.

exerted by the valve springs at full lift highest.

The Pinto camshaft and rocker wear problems associated with the standard camshaft/reground (high-performance profile) standard camshafts and standard rockers have not really been solved, it's been more a case of slowing the rate of wear to a more acceptable level. Given this problem, the use of a high lift camshaft and very high valve spring pressure, in conjunction with rockers in a non-standard full lift position, is looking for trouble. The rocker geometry has largely been the real problem with the

SPEEDPRO SERIES

Pinto engine and, more specifically, the fact that when a camshaft change was made the geometry was not restored to standard.

If a rocker and lobe fails prematurely after being set correctly the cause of the failure, provided oil is squirting from the spray bar, is that the rocker or camshaft lobe, or both, were not of the required hardness value.

VALVE STEM HEIGHT

Irrespective of the camshaft lobe dimensions and the height of the tops of the valves, every rocker can be individually repositioned for optimum geometry. This does not guarantee that a lobe or rocker will not wear out prematurely, it just means that poor rocker geometry will not be the cause of the failure.

There are two scenarios that are to be avoided as they reduce the reliability of the valve train, do not produce more power and cause 'harshness' (irregular mechanical noise) which is a clear sign of something being wrong and unacceptably rapid component wear.

The first example is when the valve stem is too short. With a short valve stem the ball-headed adjustment screw can be wound up to achieve the correct valve clearance but the geometry is, of course, no longer correct. In such a case, the valve stem needs to be 'lengthened' by the use of lash caps. Note that the stems of standard valves virtually never need to be reduced in length.

The valve stem lengths effectively become too short when the camshaft's base circle is reduced. When the valve stem is too short, and the wiping pad is situated on the rocker towards the centre of the rocker, the actual valve lift will be maximized but to use a valve which is too short to achieve this effect is folly, especially if high valve spring pressures are going to be used to achieve high rpm. The way to get full valve lift is *not* by altering the rocker geometry from the standard, but, instead, to get a higher lift camshaft and and reset the geometry of the rocker to standard specification in the full lift position.

The second scenario is where the valve stem is effectively too long. This can happen when large diameter valves (Group 1 sized) which have long stems are fitted to the cylinder head and the camshaft to be used is either standard or has a near standard base circle ('mild' camshaft). These particular valves were originally designed at this length to correct the rocker geometry when used in conjunction with a Group 1 camshaft (smaller base circle) fitted to a 2000 cylinder head.

The problem of overlong valve stems only comes about when Group 1 valves are used with a camshaft with a bigger base circle than that for which they were designed. However, valve stem length can be reduced by grinding (on a valve face grinding machine) to suit the particular application.

ALTERED ROCKER GEOMETRY

There are several problems that must be attended to when the standard camshaft is changed for one with more duration and lift. The standard camshaft lobes are ground to specific dimensions, the rockers are made to specific dimensions and the valve stems are made a certain length so that they are a specific height in relation to the camshaft when installed in the head.

This means that if any of these three components is altered the standard geometry is lost and so, to avoid valve train problems, the geometry will have to be checked and reset as necessary. This process is not as difficult as it sounds, although it does require some work to check each possible combination of components, including lash caps if they are to be used. Note that components in a range of sizes can be bought over the counter from high-performance engine parts suppliers – particularly Pinto specialists. Because of the variations possible within one engine, it's better to make up individual lash caps to the precise thickness required (within 0.010in/0.25mm).

When a camshaft is reground the camshaft lobes' base circles are reduced from the standard diameter of approximately 30.4mm/1.195in to anything down to 27.0mm/1.060in (depends on the new profile's lift and duration, but consider 26.8mm/1.055in to be the absolute minimum).

Caution! Camshafts with base circles at the smaller end of the range *must* not be used unless modifications are made to restore standard rocker geometry. Otherwise, the rocker's

Rocker (left) has been used with a valve stem that was too short. Rocker (centre) shows correct wear area. Rocker (right) has been used with a valve stem that was too long (note that wear areas are blacked out for clarity).

ROCKERS & ROCKER GEOMETRY

Excessively worn rocker pad as the result of bad geometry.

screw adjuster will have to be wound up considerably for the rocker to obtain the correct clearance and the attitude of the rocker will be well out from the standard position. This is *not* a viable proposition because the effective wiping pattern of the camshaft lobe over the rocker pad, if not off the pad, will be very short and towards the back (pivot end) of the rocker. Expect trouble!

LASH CAPS

If a high-performance camshaft with a smaller lobe base circle than the standard cam is fitted, lash caps will have to be used to correct the rocker geometry.

Lash caps (readily available in a number of sizes, though custom made is best) effectively lengthen the stem of the valve which raises the height of the front of the rocker and restores the attitude of the rocker to standard at full lift. Note that there is considerable variation in the size of rockers – not between rockers of the same type so much, but between rockers of various makes.

Any conventional camshaft profile and *any* set of rockers can be set to operate correctly. Any standard length valve can be effectively lengthened by using a lash cap (the required thickness/depth of the lash cap initially being unknown) to restore the rocker geometry to standard.

If a non-standard long stem Group 1 valve (easily identifiable by measuring), for example, is fitted into a cylinder head and proves too long for the particular camshaft and rocker combination, the only solution is to shorten the stem. Because the amount to be removed will be quite small (not usually more than 0.25-0.75mm), this is done by grinding the end of the valve down on a valve refacing machine.

Note that the term 'lash cap' is not technically correct, but it is widely used. (The original 'lash caps' were fitted to the valve stems of US pushrod V8 engines using 'rail rockers' which had a tendency to chew out the tops of the valve stems. No height alteration was affected when 'lash caps' were fitted to such engines.)

Rockers can chew out the top of the valve on a Pinto engine, too, and to stop this happening 'lash caps', as they are commonly called, serve two worthwhile purposes in this application.

A valve train that has been checked and had the geometry

Right. Lash cap fitted into the slot of a rocker just as it would be in an assembled head. Left. This lash cap has been turned 90 degrees to show side machining.

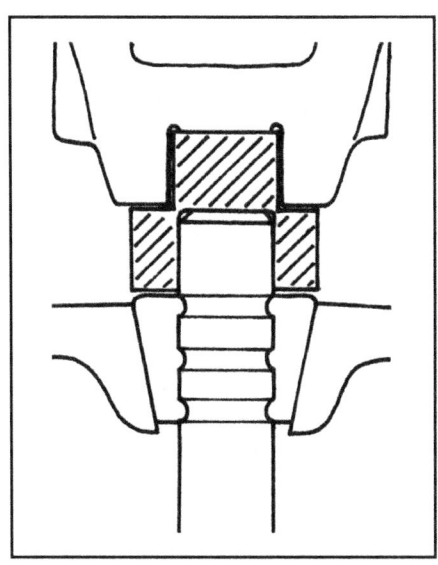

Lash cap shown sectioned. Lash cap is a neat fit over the valve stem (0.001in/0.025mm clearance).

corrected will be quiet in operation, will require less torque to turn the camshaft, will not wear out prematurely and will be more reliable.

An ideal material to make 'lash caps' out of is high carbon high chrome steel. This steel can be air hardened after the 'lash cap' has been made. The steel is heated up with a oxy-acetylene torch to bright red and then simply allowed to cool in air. The finished article will be hard but not too brittle.

The 'lash caps' must fit very neatly (diameter wise) on top of the existing valve stem and for as great a depth as possible (but not touch the tops of the keepers. A precision engineering works will make special lash caps like this for a not unreasonable amount of money.

ROCKER SIZES/DESIGNS

Rockers vary in size: not so much between the same type or design, but between the various alternative types available as replacements for standard

SPEEDPRO SERIES

items. This includes the later Ford rockers and the various 'long pad' rockers as made by some camshaft companies. In all cases the pivot radius and position on the rocker is the same overall, and so is the overall position of the rails that fit over the valve stem. The effective height of the rocker pad in relation to these two points does, however, vary by up to 1.3mm/0.050in.

Also, on some long pad rockers, the position of the centre of the pad radius in relation to the rocker varies (moved forward, which drops the pad radius down near the pivot). This lifting and moving of the pad radius centre (amount depends on the brand of rocker) is a design feature to avoid having to fit lash caps, but the range of camshaft lobe sizes and base circle sizes makes it virtually impossible to make rockers an ideal size suitable for all applications. Admittedly, any increase in the comparative height of the pad on the rocker is an improvement over the stock rocker when dealing with small base circle lobes, but this is seldom enough in itself to achieve correct geometry and lash caps will most likely still be required for perfect geometry.

The standard rockers are suitable for use with high-performance camshafts with lifts of up to 12.7mm/0.500in (measured true valve lift) and for when the full lift spring pressure is not more than 180-185 pounds.

For any given valve spring, the higher the camshaft lift the higher the pressure between the camshaft lobe and the rocker pad. It is this feature, coupled with incorrect rocker geometry, that causes the Pinto valve train to become unreliable and not, necessarily, lubrication problems.

For high-performance engines with high lift camshafts (not more than 13.0mm/0.510in) which will be consistently run up to, and sometimes over, 8500rpm, the valve springs must give approximately 60 to 80 pounds seated pressure and a maximum full lift pressure of 200 pounds. The rockers to use in such applications are

Long pad rocker which has the pad nearer the pivot compared to a standard rocker.

Late standard rocker from a high mileage engine but which has been used with correct geometry and an 'over the nose' spring pressure of 175 pounds.

those that have hardened segments fitted to them and this type is available from Crane or Emerald, to name but two suppliers. The original rockers are hardened, but they are not as hard as two-piece aftermarket rockers. Lash caps of the design shown should be used to protect the tops of the valve stems from excessive wear.

As a point of interest, Datsun engines of the 70s and 80s used a very similar rocker arrangement to the Pinto. The Datsun engines had hardened segment rockers fitted to them as standard and were absolutely trouble-free, even in the most rigorous racing situations. The spray bar which fed oil to the rockers did not always (by design) have a hole for direct spraying of oil on to each rocker. So some rockers received oiling by splash from the adjacent rocker!

Note that virtually all high-performance camshafts do not lift the valve the amount claimed by the manufacturer. The rocker ratio plays a part here but, irrespective of this, if the lift is found to be down from that specified, but the attitude of the rockers is correct at full lift, don't worry about the loss of lift. The Group 1 camshaft, for instance, is rated at 12.26mm/0.482in full lift (inlet) but will frequently be found to give

Holbay forged and machined roller rocker assembly.

ROCKERS & ROCKER GEOMETRY

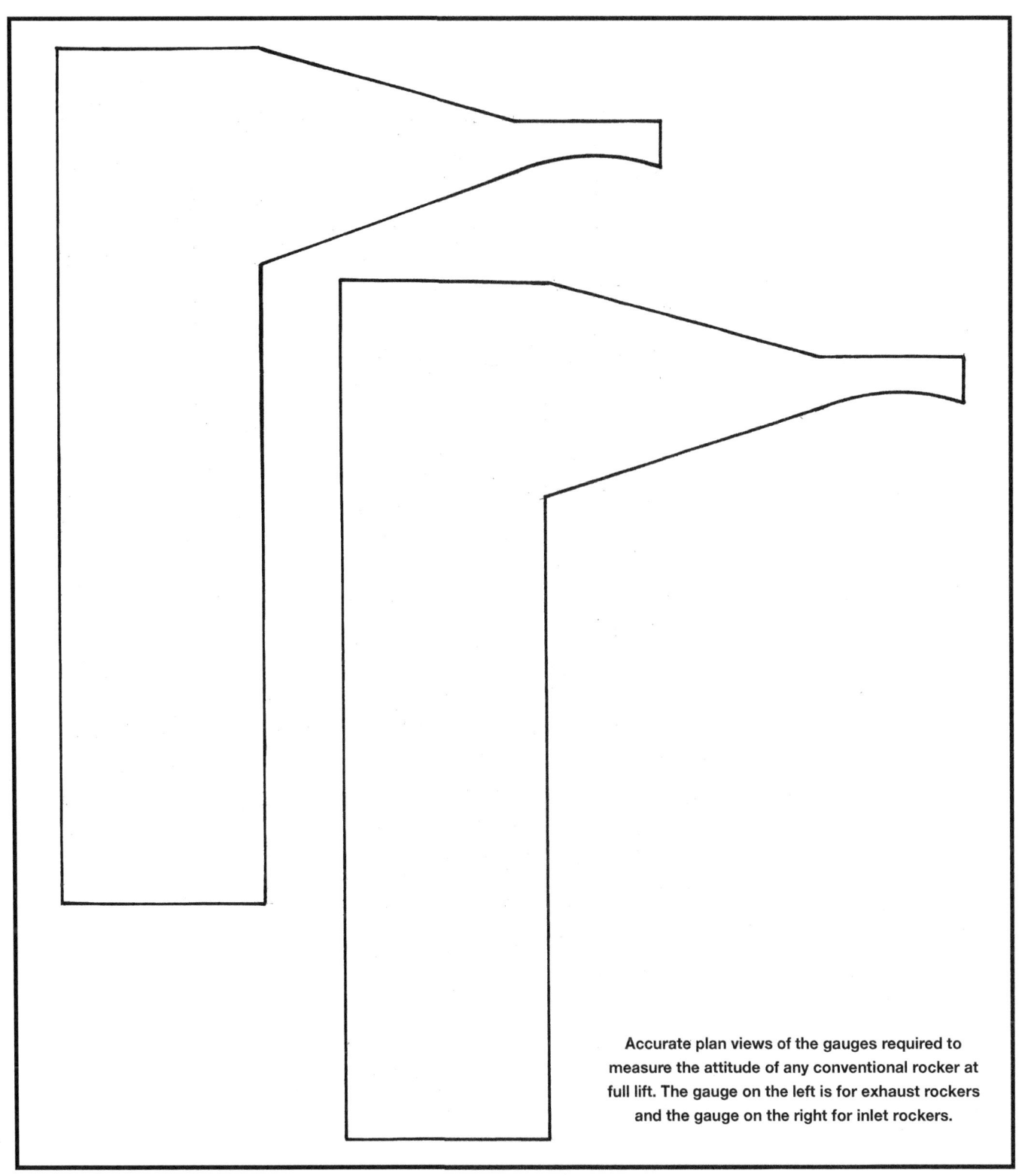

Accurate plan views of the gauges required to measure the attitude of any conventional rocker at full lift. The gauge on the left is for exhaust rockers and the gauge on the right for inlet rockers.

SPEEDPRO SERIES

11.8mm/0.464in with correct rocker geometry. The rocker attitude can be altered to obtain the advertised lift, but no increase in power results and the valve train geometry is no longer correct. *Always* run the rockers with the correct geometry for maximum reliability.

Roller rockers

Note that if 'drop in fit' roller rockers are going to be used, the rocker geometry *must* still be checked using a conventional rocker first and then substituting the roller rocker. Roller rockers are not a 'cure all' and their use does not mean that the geometry is automatically correct. The roller rockers have to be the same size as the conventional rocker that they are going to replace. If roller rockers are run with incorrect geometry, the rollers tend to 'pick up' on the centre pin and seize.

Geometry is checked by first fitting a standard rocker and setting the attitude of the rocker pad to the gauges at full lift. The correct size lash caps are then made up before the roller rockers are substituted.

CHECKING ROCKER GEOMETRY

Note that, irrespective of camshaft lift, lobe base circle diameter or the type of rocker used, the curve of the rocker pad can only be in one ideal position at full lift. The attitude of the standard rocker at full lift revolves around the curve of the rocker pad (the radius of the curve is 50mm) being equidistanced about the nose of the camshaft lobe.

There is some tolerance on the position of the curve at full lift, but not a lot. It does not take too much of an alteration of the rocker's position (achieved by raising or lowering the top of the valve stem with lash caps) to get the curve noticeably out of position. This is not something that can be guessed by looking at the rocker at full lift. It has to measured as accurately as possible and this is not easy. No claim is made that the method described here is completely accurate, only that it is accurate enough to ensure that rocker geometry is in tolerance.

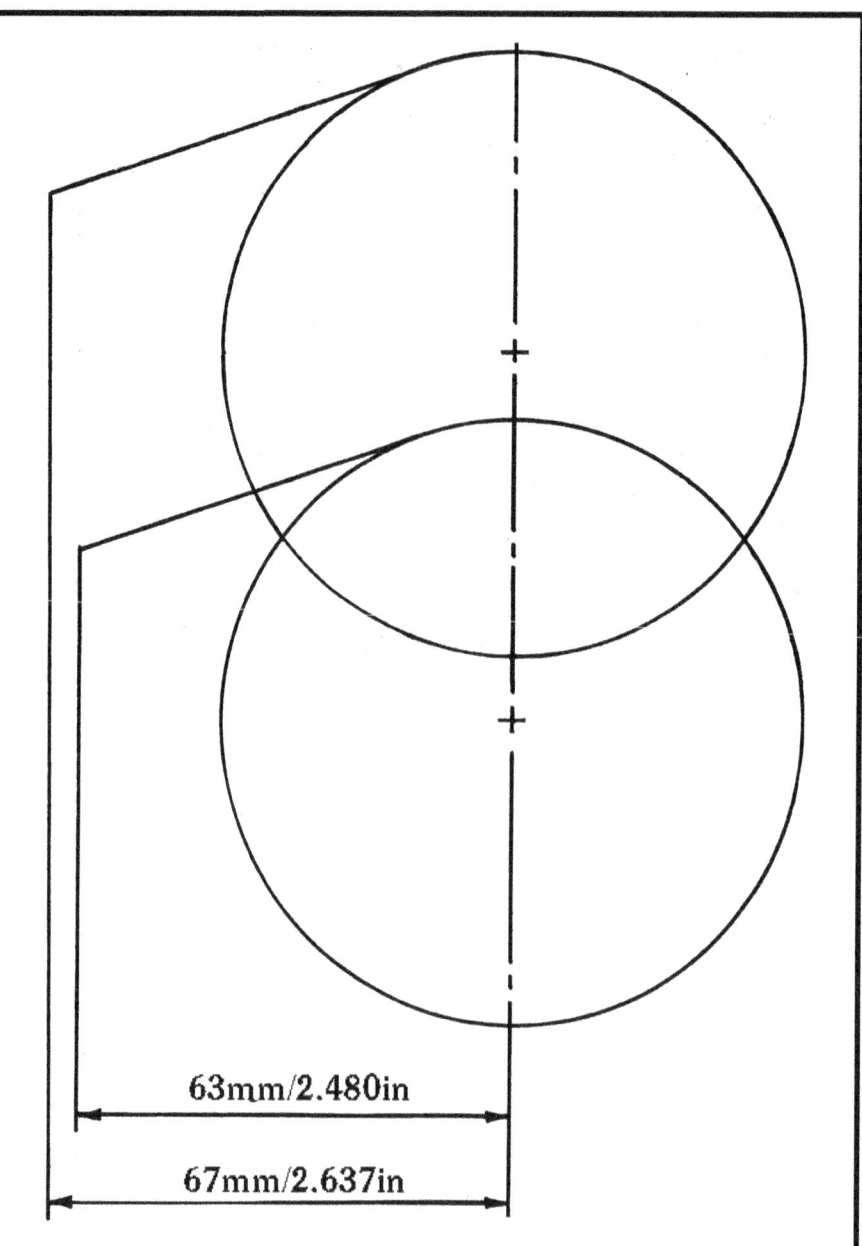

The rocker geometry gauges are based on the effective diameter of the rocker pads (85mm/3.346in) represented by the two circles, and the distance from the side of the head to the centre of the rocker pads (63mm/2.480in exhaust, 67mm/2.637in inlet).

ROCKERS & ROCKER GEOMETRY

Camshaft moved back 5mm in the cylinder head, lobe is in the full lift position and there is a shim between the valve stem and the rocker which, for checking purposes, has effectively increased the length of the valve stem.

The usual scenario will see the back of the rocker (the pivot end) too high and the pad higher on the side of the pivot at full lift. Basically, this is caused by the camshaft lobe base circle being smaller than standard and the valve stem length becoming effectively too short. Here is the cause of many a rocker failure (irrespective of how much oil is sprayed over the rocker).

The next part of the procedure is measuring the position of the arc of the rocker pad when the camshaft is at full lift. To do this two separate gauges will have to be made up as per the photographs and diagrams that accompany this text.

The gauges are made up using sheet steel (minimum thickness 1.5mm/0.062in) or 16 gauge aluminium. The dimensions are marked out on the flat sheet and

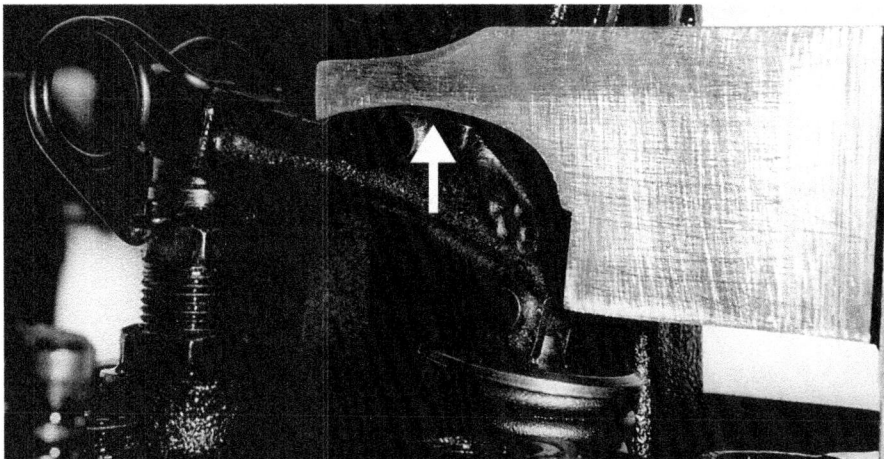

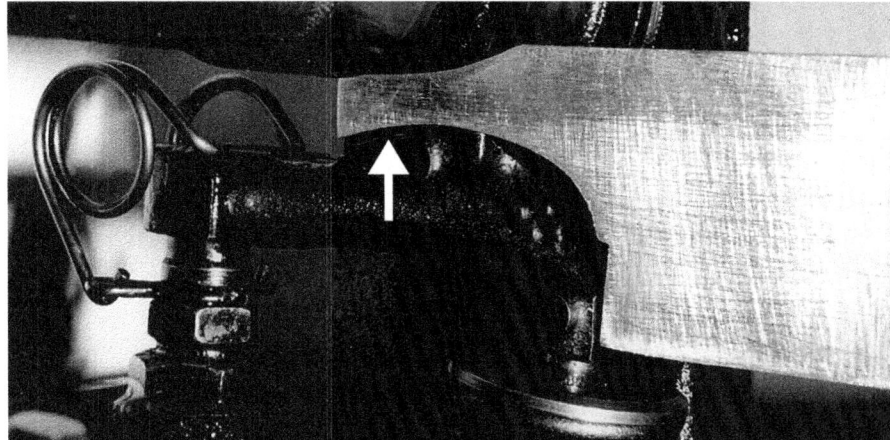

Geometry gauge in use with the rocker at full lift showing the three possible situations. Top - there is a noticeable gap at the end of the pad nearest the valve, meaning the valve stem is too short. Centre - this time the gap is at the end of the pad nearest the pivot, meaning the valve stem is too long. Above - gauge fits the pad perfectly, meaning geometry is correct.

SPEEDPRO SERIES

then cut and filed accurately to size. The finished gauges can be directly overlaid on to the diagram.

One gauge is for the inlet rockers and the other is for the exhaust rockers. The two gauges differ from each other because the measurement datum point used is the side of the cylinder head and the distance from the centre of the inlet valve rocker to the side of the head is not the same as the centre of the exhaust valve rocker to the side of the head. The sides of the cylinder head *must* be absolutely clean and free from all gasket material and sealer.

The gauges locate on the different sides of the cylinder head (inlet gauge on the inlet side and exhaust gauge on the other) and hand pressure is used to ensure that the gauge is firmly held against the side of the cylinder head. The gauges fit into the available space and a direct visual reading is taken. This may seem a bit hit and miss but, in the final analysis, is very accurate.

To gain access to each rocker, the camshaft thrust plate has to be removed and the camshaft moved back about 5mm/0.200in. Before the camshaft is moved back, to allow access for the gauge, the camshaft should be in the normal running position and the valve clearance set on the rocker. The camshaft is then pulled back 5mm/0.200in (approximately) and then turned to position the rocker at full lift. The valve springs used do not have to be strong ones; in fact, any reasonable spring tension can be used (20 pounds seated is enough).

Every valve and rocker combination *must* be checked and set up individually and only one rocker is on the cylinder head at any one time. Each rocker is set with the correct valve clearance and then turned to the full lift position. This is easy to find as the valve stops moving (opening). At this point the gauge is placed in position and a visual reading taken of the attitude of the rocker arm wear pad in relation to the curve of the gauge.

The next step is to increase the height of the valve using a combination of small flat shims the width of the slot in the rocker and about 25mm/1in long. These shims are then placed under the rocker and on top of the valve stem. The camshaft is moved to the normal position and the valve clearance is reset. The camshaft is then moved back about 5mm and turned to the full lift position.

The gauge is positioned on to the side of the cylinder head again and a visual reading taken of the attitude of the rocker's wear pad in relation to the gauge. Shims are added to the top of the valve stem until the rocker has the correct attitude. The total thickness of the shims used is the exact thickness/depth requirement of the lash cap. To get the precise thickness/depth required for each individual the lash caps may have to be custom made.

While this is tedious work, it realistically only has to be done once. The 'lash caps' will often be different thicknesses and will always have to be kept in order whenever the cylinder head is being maintained. This is all really a small price to pay to have ideal rocker geometry and reliability.

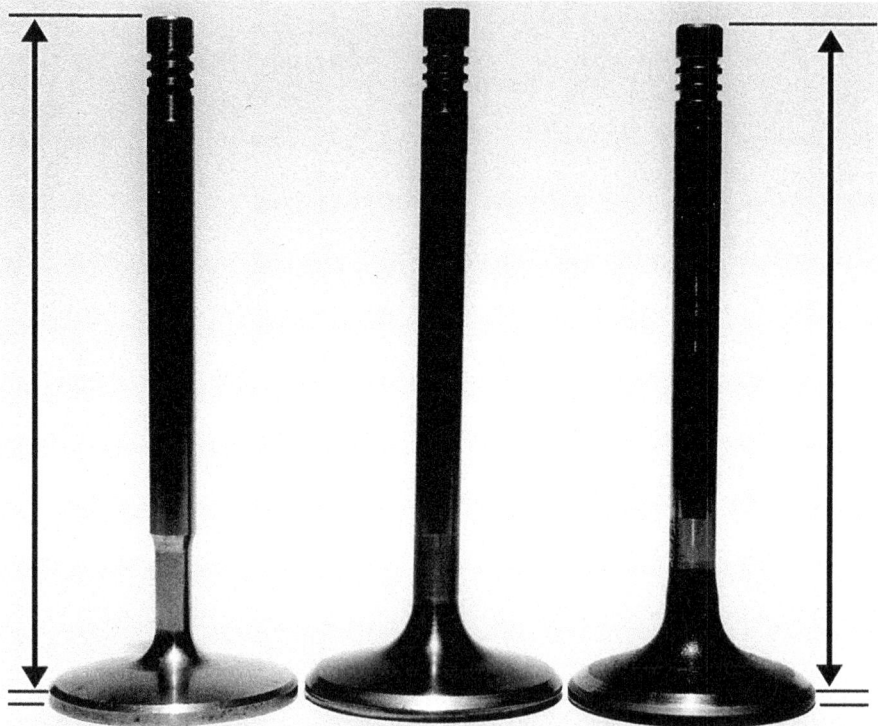

Note valve length is actually the distance from the seat to the tip of the stem and not the total length of the valve.

www.velocebooks.com / www.veloce.co.uk
All current books • New book news • Special offers • Gift vouchers

Chapter 10
Exhaust systems

STANDARD CAST IRON EXHAUST MANIFOLD

The standard cast iron exhaust manifold should be discarded in most instances (except for some classes of motorsport) and replaced by a tubular exhaust system. The standard cast iron exhaust manifold, while being very strong, is not efficient over 4000rpm.

EXHAUST SYSTEM CONSTRUCTION

Frequently optimum design (such as equal length primary pipes) has to be compromised because the installation will not allow it. When this happens, keep to the optimal criteria as much as possible and only compromise where absolutely necessary. The difference in engine performance under 6500rpm between having an exhaust system with all features optimized compared to one with, say, varying primary pipe lengths, will be minimal.

Avoid using exhaust system tubing with bends which are squashed down when formed, such bends will not have the correct internal

'Squashed' bends are restrictive and therefore not ideal for a high-performance engine.

cross-sectional area and will cause a restriction. Note that if an exhaust system is available which does have squashed bends, but the pipes are on the large side for the application, it will more than likely prove to be quite acceptable. If an exhaust system is being custom made to suit the application, insist on hydraulically bent or mandrel-formed bends which have little or no deformation.

Systems are made up of straight lengths of tube and bends of varying angles and will usually have welds at each joint: this is an entirely acceptable method of constructing an exhaust system.

Care must be taken when exhaust pipes are welded to ensure that the two pipes do line up completely when they're 'tacked.' The pipes to be welded must butt up to each other

SPEEDPRO SERIES

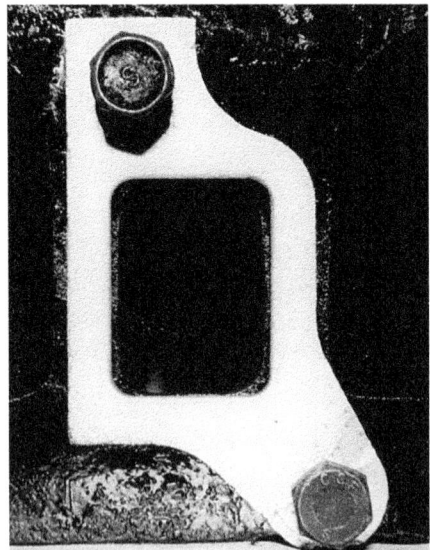

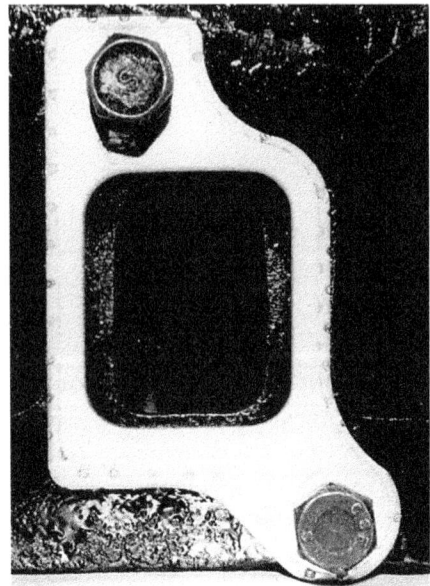

Standard exhaust gasket (top) and largest available (above).

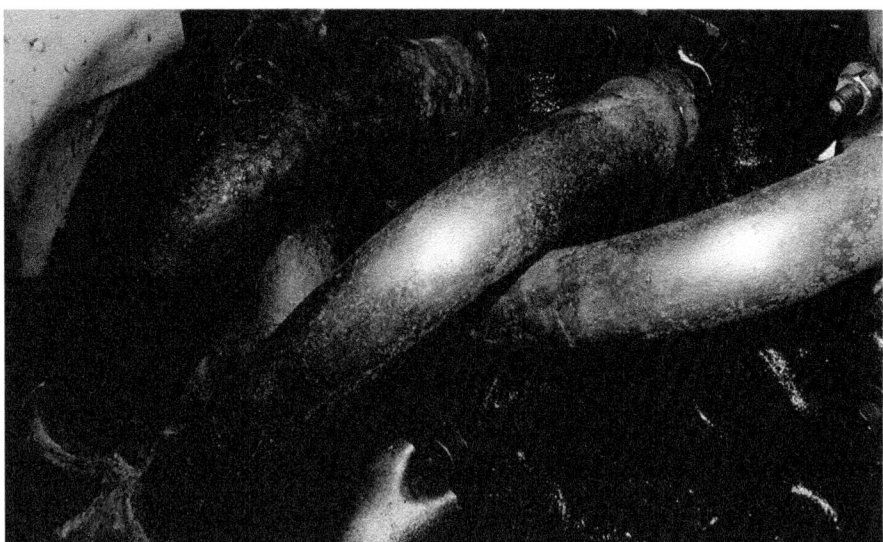

Four into two into one system with 38mm/1.5in primary pipes, 45mm/1.75in secondary pipes and a 54mm/2.125in main pipe.

with a minimal gap all round so that, when they are welded (usually using a mig welder), there will not be any 'over weld' inside the exhaust pipe. If this happens, the inside diameter of the tube will actually be reduced at the weld and this is not desirable.

Do *not* grind the welds to make them look as if there is no weld present, as strength will be reduced and the pipe may break at the weld as a consequence. It is acceptable engineering practice (for exhaust pipes anyway) to leave all welds as they are.

If you wish, stainless steel can be used in the fabrication of tubular exhausts.

FOUR INTO TWO INTO ONE

This type of system has four primary pipes converging into two secondary pipes and these two pipes converging into one pipe. The pipe diameters and lengths will vary depending on the installation space available, but the overall pattern of all systems should be similar.

The primary pipes will usually be 38mm/1.5in or 42mm/1.625in or (possibly 45mm/1.75in) outside diameter and be approximately 305mm/12in to 457mm/18in long; the secondary pipes are 42mm/1.625in or 45mm/1.75in (possibly 51mm/2in) outside diameter and anything from 508mm/20in to 610mm/24in long; the main pipe is of 51mm/2in or 54mm/2.125in outside diameter. Such a system is excellent for low to mid-range power (road use or competition) but, while it works well right through the rpm range, it does not normally produce the top end power that a four into one system produces.

FOUR INTO ONE

This is the most common configuration for racing purposes and nearly always produces better top end power than a 4 into 2 into 1 exhaust system. When smaller diameter primary pipes such as 42mm/1.625in are used in conjunction with a 51mm/2in main pipe, it will prove very suitable for all-round use (low, mid and top end performance). 38mm/1.5in primary pipes are the ideal size for 1600 engines but 42mm/1.625in diameter primary pipes will still work well.

PRIMARY PIPES
Primary pipe diameters
For 2000 engines the primary pipe diameter for best mid-range and

EXHAUST SYSTEMS

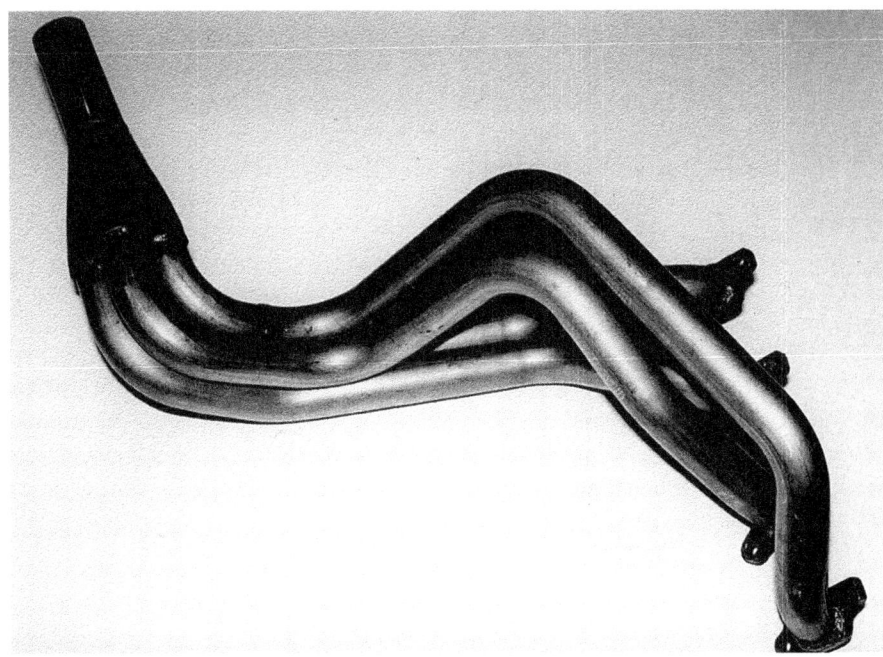

Four into one exhaust system with 42mm/1.625in primary pipes and a 51mm/2in main pipe. The primary pipe lengths are all different!

low end power (all road use engines) is 42mm/1.625in for either the 4 into 2 into 1 or the 4 into 1 exhaust configuration.

The best diameter for top end power (motorsport) for a 4 into 1 system will be 45mm/1.75in or 48mm/1.875in outside diameter. Stay away from the larger primary pipe size unless the engine is being regularly run to 8000rpm, or more.

Note that a well tuned engine with 38mm/1.5in primary pipes (the smallest ever to consider on any Pinto 1600, 1800 or 2000) will still run to approximately 7000rpm on a 2000 engine, but it will not have 'urgency' in getting to those revs, especially above 6500rpm. Lower down the range (2000 to 6500rpm) performance will be better than with a large primary pipe. Don't be blinded by size, tune for the application! By way of explanation, the large diameter primary pipes raise the point in the rpm range where the peak torque will be. They do *not* increase the amount of torque produced by the engine, they only shift the point at which peak torque occurs. For example, a well modified engine may develop peak torque at 5000rpm when using 42mm/1.625in diameter primary pipes with a 54mm/2.125in diameter main pipe. The same system with larger 48mm/1.875in outside diameter primary pipes and a 64mm/2.5in main pipe might well change the point of peak torque to 5800rpm with a corresponding power loss at low rpm. The engine will feel more 'urgent' in the upper rpm range because it will have more torque (or turning force) at those revs. Correspondingly, the engine will have less torque lower down in the rpm range, and quite noticeably so. More gear work will be necessary to compensate. To use larger primary pipes effectively, the engine rpm has to be kept in the correct range. Once again only fit this type of exhaust system if high rpm (competition) is being used most of the time.

Primary pipe lengths

The shortest primary pipe length used is approximately 610mm/24in long and the longest used approximately 915mm/36in long. The longer pipes tend to be better but, sometimes, it's just not possible to fit them. Primary pipes are usually made to the short sizes or the long sizes and not in between, even though this would be perfectly satisfactory.

Frequently, a 4 into 1 system will have varying primary pipe lengths because of space limitations. For example, the primary pipes could be 660mm/26in, 736mm/29in, 813mm/32in and 915mm/36in to fit everything in but overall efficiency won't be reduced to any marked degree because of this difference in pipe lengths.

Equal length primary pipes

When making an exhaust system it is ideal to have all of the primary pipes exactly the same length. Frequently this is not possible but engine power, in most instances, will not be reduced because of it. Under 7000rpm there will certainly only be minimal losses, even if every pipe is a different length. It can be quite difficult to actually make all pipes the same length and make the configuration look totally presentable. In many instances the ideal of having all primary pipes the exact same length is sacrificed to make the pipes fit well and look right in the engine bay.

There is no doubt that a correctly designed and built exhaust system made specifically for a particular application will enable an engine to produce top power. However, off the shelf systems are often not far behind in efficiency.

Tubular exhaust pipe manufacturers make systems according to demand and this means that the 2000 engine is better catered for than the 1600/1800. The pipe diameters of readily available systems are usually larger than recommended for the 1600 engine in particular. For the 1600/1800 with a 4 into 2 into 1 system, consider 38mm/1.5in primary pipes, 42mm/1.625in secondary pipes and a 48mm/1.875in main pipe to be ideal for all round use. For a 1600/1800 4 into 1 system, consider 38mm/1.5in primary pipes (610-710mm/24-28in long) leading into a 48mm/1.875in main pipe to be ideal for all round use. There are a quite a number of commercially available tubular exhaust manifolds available from a number of sources to suit just about every application imagineable.

All road going engines must be well silenced these days and the use of virtually any large bore (at least the same pipe diameter as the main pipe of the extractor) silencer will be acceptable. Most 'universal fitting' silencers are not all that restrictive. For racing applications, most venues insist on effective silencing (95dB) and silencers of the straight through variety are the most commonly used but some of them can be marginal. Large pipe diameter, aluminium case or carbon-fibre case silencers are available but they are quite expensive.

If cost is a major consideration, use a 'universal fitting' baffle type silencer which will not be expensive at all. 'Universal fitting' type silencers are available from specialist exhaust shops who always have in stock a range of silencers for all sorts of fitting applications. The main criteria with these silencers is the space available for fitting the silencer (they are usually oval) and that the pipe diameter of the silencer is preferably $1/8$ of an inch larger than the main pipe diameter of the exhaust system.

Chapter 11
Flywheel & clutch. Engine balance

FLYWHEEL

The standard flywheel is very heavy and, while this is fine for the standard road-going vehicle, it's less than desirable for a high-performance engine. Lightening the flywheel of a modified engine is definitely advantageous and the engine will be much more responsive because of it.

Removing material from the outer diameter of the flywheel is more beneficial than removing material from anywhere near the centre (force times distance). **Caution!** Flywheels *must* not be machined too thin (no material section should ever be less than 8mm/0.312in) as this may lead to breakage which could be dangerous. **Caution!** New flywheel bolts, coated with a locking agent (Loctite or similar) and torqued to the factory recommended maximum *must* always be fitted when rebuilding an engine for high-performance work. Note that using a locking agent on the flywheel bolts does not guarantee that the bolts will not come undone.

There are six flywheel retaining bolts on the standard crankshaft but no dowels. All flywheels of high-performance engines should have at least one dowel.

The standard 2000 weighs about 6.5kg/14 pounds and can be lightened by turning on a lathe, without detriment to strength, to about 4 kg/9 pounds. The 2000 flywheel is recommended for use on the 1600/1800 engines because of its larger clutch (a lightened 2000 flywheel is lighter than the standard 1600/1800 flywheel). Very light and strong aftermarket flywheels are available, but the standard flywheel when lightened and then perfectly balanced is quite satisfactory for use up to 7500rpm but, for safety sake, any engine that is turning more than 7000rpm on a regular basis should be fitted with a lightweight steel flywheel assembly. Racing being what it is, of

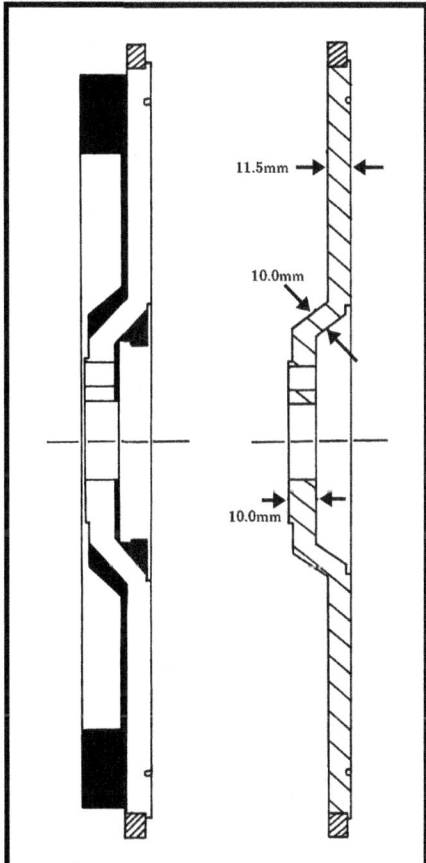

Cross-section of standard Pinto flywheel. Black areas of diagram on left indicate material to be removed. Diagram on right shows minimum material sections.

SPEEDPRO SERIES

Top - the rear (block side) of a lightened standard Pinto flywheel. Above - the front (clutch side) of the same lightened flywheel.

Twin plate AP racing clutch on lightweight aftermarket flywheel.

course, anything can happen. Higher rpm than you think you will use could end up being used more frequently than desirable. Overspeeding the engine even once could be once too often! Don't take this sort of risk as it is just not worth it. All standard Pinto flywheels can be lightened.

To make a worthwhile reduction in weight, material only needs to be removed from the back of the flywheel (faces the block). The clutch side of any used flywheel *must* be cleaned and trued which involves removing 0.25-0.50mm/0.010-0.020in from the clutch plate and pressure plate location surfaces. Pinto flywheels are quite tough to machine, so the removal of the necessary material will take considerable time.

Note that the surface on which the clutch plate 'runs' will have to be ground true if it shows signs of having been overheated, as these heat marked surfaces are very hard and not easy to turn using a tungsten carbide tool. It's better to grind the surface because a turned surface, while looking quite good, will usually have high spots right where the overheating took place.

Burton Power sell 5 lightweight (9 to 12 pound) steel flywheels for six bolt retention SOHC Pinto engines and 9 bolt retention Cosworth crankshaft equipped engines. They sell AP Racing up-rated pressure plates and clutch driven plates of either organic lining material or cerametallic linings on paddle clutch plates. The AP Racing cerametallic clutch assemblies are

FLYWHEEL & CLUTCH. ENGINE BALANCE

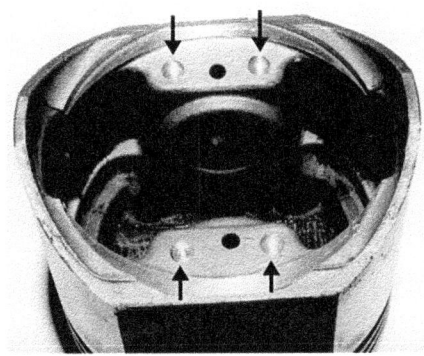

Underside of standard piston that has had material removed (arrowed) from the piston pin bosses to balance weight.

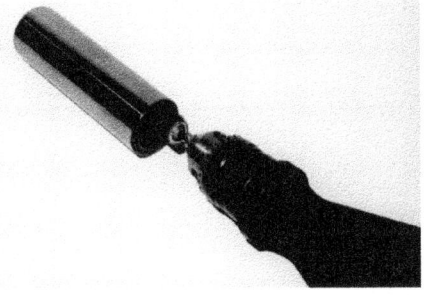

Piston pin having material removed from the inside using a high-speed grinder fitted with a mounted point grinding wheel. Up to 5 grams can be removed easily using this method without detriment to the strength of the piston pin.

virtually indestructable no matter how much abuse the clutch is given. They are essential for rigourous racing applications. AP Racing twin plate organic lined clutches are also available from Burton Power.

Holbay Engineering make aluminium and steel flywheels to order for any combination. Their high tensile aluminium flywheels have an EN8B steel segment 8mm thick let into the surface of the flywheel for the clutch plate to work against. They fit AP Racing and Sachs clutch assemblies.

ENGINE BALANCE

Ford engines have always been noted for their good machining and balance and the Pinto engine is no exception. On manufacture all Pinto engines were balanced to within the factory limits and, apart from the odd exception, will prove to be well enough balanced for just about any application using engine speeds up to 7000rpm. For all road-going engines the standard factory balance is quite adequate, although some engines are better balanced than others. The connecting rods and pistons are quite accurately balanced but the factory balance tolerance for the crankshaft is wide so it may be found to be slightly out of balance and need attention. For any competition application the crankshaft *must* be checked, and adjusted as necessary, to ensure that it is perfectly balanced. A properly balanced crankshaft will take 8500-9000rpm reliably.

For engines which will rev over 7000rpm and all competition applications, the engine *must* be rebalanced to ensure that all the relevant components are 100 per cent in balance. This process will require the crankshaft to be spun on a dynamic balancer and rebalanced if necessary.

Next, the flywheel (completely new/lightened original) is bolted onto the crankshaft, as is the crankshaft pulley/damper and the assembly balanced as a unit. A new clutch pressure plate is then bolted on to the flywheel and the whole assembly balanced as a unit. The crankshaft, flywheel and new pressure plate *must* all have their relative positions marked so that, if they are ever dismantled, they can be reassembled in exactly the same positions as when they were balanced.

The connecting rods are compared end for end, then balanced on a purpose built machine, of which there are several types. Some of these machines give weights in grams but other weighing devices don't give weight readings and balance is achieved by comparison. In all cases the lightest small end is the standard against which the other three connecting rods are matched by the removal of material. Similarly, the lightest big end is the standard to which the other three big ends are matched. Whatever the method used, in the hands of a skilled operator the connecting rods will end up perfectly balanced. The original connecting rods will almost always have been balanced by Ford to within 2 to 3 grams.

Each piston and its piston pin are weighed together (variation is seldom more than 3gm/0.10oz). With the lightest piston and pin as the standard, all other pistons must be lightened so that they are all within 1gm/0.035oz of the standard. Piston rings are not weighed as they are always the same weight within the same set.

Piston material can be removed from the piston pin bosses or the underside of the crown adjacent to the oil ring. The inside of the piston pin can also be lightened using a high-speed grinder and a suitable mounted point. As the piston and pin are weighed collectively, material can be taken from either component.

www.velocebooks.com / www.veloce.co.uk
All current books • New book news • Special offers • Gift vouchers

Chapter 12
Ignition system

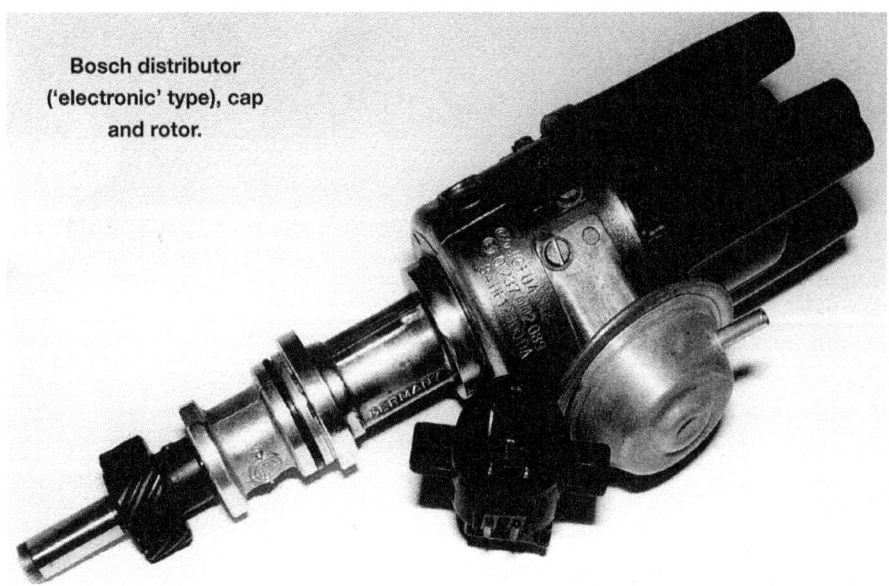

Bosch distributor ('electronic' type), cap and rotor.

Please note that in-depth information on building a high-performance ignition system, distributor modifications and optimised ignition timing can be found in another Veloce Publishing SpeedPro Series book – *How To Build & Power Tune Distributor-type Ignition Systems* by Des Hammill.

Either type of original equipment Bosch distributor (G-FU-4 and J-FU-4) is ideal for use on the Pinto engine (in electronic or contact breaker form). The electronic distributor will use a higher output coil, or one that is rated for electronic ignition systems. The contact breaker points type ignition can use a standard ignition coil but, ideally, will use an uprated low voltage sports type coil (a high-voltage electronic coil will burn the points out if is used on a points type distributor).

There are certain requirements which simply must be attended to if the ignition system is going to be reliable and consistently produce the quality of spark necessary to successfully run a high-performance engine. If any of the components mentioned in the following sections is not in optimal condition, then ignition system performance will not be as good as it could be.

Mention is made of OEM (original equipment manufacturer) parts and the use of them. If the parts source available to you does not deal with OEM parts and only offers replacement parts from alternative manufacturers, use the non-original parts but check them in the manner described. The inference is not that alternative replacement parts are inferior to original equipment, it's just that original equipment parts are specifically rated for the distributor concerned. The

IGNITION SYSTEM

alternative replacement may well be rated differently, and therefore be slightly less suitable for the specific application, than the original equipment part or parts. Irrespective of who makes the distributor parts they should all be new or near new. Fit new contact breaker points frequently in high-performance applications.

DISTRIBUTOR SPINDLE

The fit between the spindle of the distributor and the bearing in the distributor body is of vital importance if the distributor is of the contact breaker points type. If there is any sideways movement at all (0.001in maximum clearance), the distributor

Distributor drive gear.

Distributor spindle and cam plate as a unit.

has obviously been well used and is not suitable for any performance application. Such a distributor will have to be reconditioned, which will mean that the spindle bearings and the spindle will be replaced. A less expensive option is to go to a breaker's yard and buy an identical distributor which does not have any sideways spindle movement. This could mean that the distributor is actually off a much later engine or one with lower mileage.

The reason spindle to bearing fit is so critical is that, as the rotational speed of the spindle increases, the spindle does not follow the central path. On well worn distributors, the spindle gyrates around and the gap of the points increases, altering timing and dwell. No high-performance engine can operate efficiently with a worn distributor spindle and/or bushes.

Some distributors are notorious for wear in this area, while other distributors that fit the same engine never seem to wear. All distributors are good when new. Take the trouble to get a good distributor body with sound spindle and bushes as it is the only basis on which to build a good ignition system.

If the spindle and/or the bush/es have any wear, have the distributor reconditioned with new parts or find an alternative distributor which is in excellent condition and perfectly serviceable.

DRIVE GEAR

This gear is driven off the auxiliary shaft and will not normally show too much sign of wear. Replace any gear with teeth that are knife-edged and look like they are worn. If in doubt about whether the gear is worn or not, check the original gear against a new one. Gears with misshapen

Shims used to control spindle endfloat.

(worn) teeth have too much backlash and timing fluctuations are a possible consequence.

ENDFLOAT/ENDPLAY

The distributor spindle's endfloat/endplay/lash is usually controlled by the amount of clearance between the drive gear and the body of the distributor. The workshop manual for your engine lists the minimum and maximum amount of endfloat permissible. If possible, set the endfloat to the minimum amount recommended by the manufacturer.

CONTACT BREAKER POINTS

As a general rule, the original equipment manufacturer usually makes the best set of points for a given distributor. However, some pattern parts are just as good as OEM parts and some are better than the originals. For example, some pattern parts feature hollow contacts and a separate current conducting strap and are all-round first class components, and at a reasonable price. High-performance engines should only have the best quality points sets fitted, irrespective of the cost or manufacturer who makes them. Fit

SPEEDPRO SERIES

Typical set of contact breaker points.

a new set of top quality points to the distributor and always carry a spare set. Note that it's inadvisable to push a metal feeler gauge through the point contacts as this can contaminate them. Place the feeler gauge next to the contacts to set the gap or use a piece of clean cardboard of the right thickness (micrometer measured) in place of a metal feeler gauge. Use the minimum recommended gap.

CONDENSER

Fit the correctly rated unit, preferably an OEM part. An under- or over-capacity condenser burns one side of the points more than the other side and a faulty condenser causes the engine to misfire. An important thing to check always is that the condenser is securely screwed to the distributor: a loose condenser will cause the engine to misfire erratically!

ELECTRONIC MODULE

The right module, correctly rated for the particular distributor, must always be fitted. Modules are frequently mounted on an aluminum plate (which acts as a heat sink) which is in turn bolted to the body (firewall/bulkhead or inner wing).

The electronic module requires special equipment to check its operation. If a module is suspected of being faulty (engine misfire, no spark) it has to be tested or an alternative module fitted and the engine run.

Typical condenser.

Modules are usually expensive and testing the original is more practical. The module tester connects to the module and puts it through a cycle that starts at the simulated idle speed and takes it up to simulated full speed. Modules do fail, and they fail far more frequently when subjected to excessive heat and vibration. It's a good idea to carry a spare. Many garages now have excellent test equipment on site and can check modules quickly and easily.

DISTRIBUTOR CAP

Fit a new cap to the distributor and protect it from damage when it is off the distributor (when replacing points, for example). Wrap the cap in a clean rag with the leads attached and place the cap out of harm's way. Avoid scratching the cap or knocking

Typical electronic module.

Distributor cap with horizontal HT terminals. Ideal for carburetted Pintos.

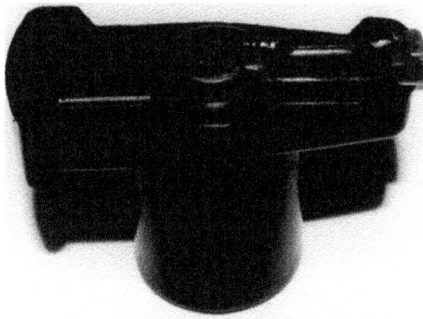

Typical standard rotor arm.

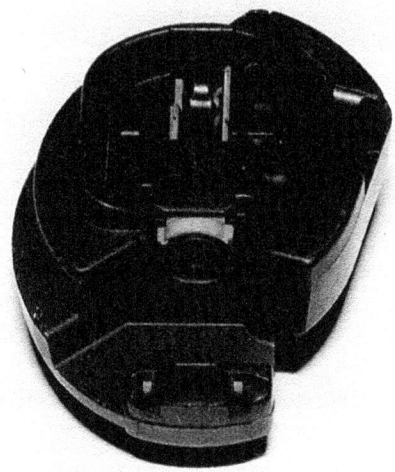

Bosch governor rotor – this one is rated at 6200rpm. The rpm at which the rotor cuts the ignition is marked on the unit's underside.

IGNITION SYSTEM

the cap as this can lead to cracks and a cracked cap will cause an electrical failure (engine misfire). Some manufacturers use copper contacts and some use aluminum: copper being the better of the two. Caps from Bosch are of excellent quality and are very robust but they do cost a bit more.

ROTOR ARM

Fit a new rotor arm to the distributor. These items are relatively inexpensive and new units are usually trouble-free. They must be a tight push-fit on to the distributor spindle; it's no use having a rotor arm that flops around on the spindle. When the rotor is loose the problem can be with the spindle or the rotor arm or both. Check the fit of the rotor on to the distributor spindle and improve the fit, if necessary, by wrapping the spindle with thin tape. The tape cannot always be wrapped around the spindle for the full 360 degrees because of the spindle slot, but it can be wrapped slot edge to slot edge around the spindle.

COIL

A wide range of coils is available and some of them, while being suitable for electronic distributors, are not suitable for points type distributors. Coils for the types of ignition systems under discussion here can be divided into four basic groups.

Low voltage conventional (standard) coil

The conventional standard 12 volt ignition coil suitable for use with points distributors will have an output of approximately 17 KVs plus. This type of coil is regarded as a low voltage type, but in no sense does that mean inferior. These coils are generally far better than they are given credit

Clockwise from top left - standard conventional coil; typical uprated coil (approx. 28Kv); typical balast resistor coil; typical oil-filled 'electronic' coil (up to 37Kv).

for and their replacement with an alternative 'high-performance' coil does not automatically mean more power and efficiency. There are many other things to consider – such as plug wires (low resistance type), for example – before replacing a standard type coil with an uprated coil.

The majority of points type ignition systems will run very well with this sort of coil provided the engine's compression is not too high (over 10:1), the rpm that the engine is turned to is not too high (up to 6000rpm) and the cylinder head/s are reasonably standard. When engines are uprated with worked heads (improved volumetric efficiency), increased compression (higher cylinder pressures) and high rpm operation, the situation changes. Having more air/fuel mixture inside the engine to compress, higher cylinder pressures and often needing more sparks per minute all add up to more coil spark being required.

Uprated low voltage conventional coil

Uprated coils (Lucas Sports or Accel, for example) are available which, while still being of the low voltage type, are rated at approximately 26 to 28 KVs. These coils are very similar in construction to standard coils but, because of subtle internal differences, produce more secondary voltage. These coils can be used as direct replacements for standard ignition coils and are quite satisfactory for use with all points type ignition systems. This type of coil will not burn the points out and does have more output than the standard factory-fitted coil.

These coils are suitable for 7500 to 8000rpm operation. Once an engine has been modified (more compression, head work, camshaft, higher rpm being used) the fitting of one of these coils is recommended. Improved ignition performance (over a standard type coil) is only possible if all other related components (plugs, points, plug wires and so on) are in perfect condition.

Ballast resistor coils

When the ballast resistor is in circuit, these are low voltage coils and will not burn the points out. They are typically 9 volt coils which have the 12 volts of the electrical system fed to them only for starting, during which time they are high voltage coils (but if run continuously like this, they would burn the points out). Once the ignition key is let go after the engine has started, the 12 volts of the electrical system

SPEEDPRO SERIES

is passed through the ballast resistor which reduces the primary voltage to the coil's rated voltage. The ballast resistor has to be correctly rated to suit the particular coil to ensure the coil has the correct voltage during normal operation. This coil-type is used by car manufacturers to provide improved starting and there's no doubt that engines equipped with this system are getting a very good spark.

Low Voltage Ballast Resistor Coil – 17KV plus (approximately) – If a ballast resistor coil is fitted to an ignition system without the ballast resistor (becomes a high voltage coil) the points will burn out fairly quickly. The idea of putting 12 volts across a 9 volt coil to obtain a better spark is well founded but, as the points will burn out rapidly, it cannot be done for very long.

High Output Ballast Resistor Coil 35KV (approximately) – Accel and Mallory manufacture excellent ballast resistor coils suitable for any ignition system, even though these coils are primarily made for their own dual point distributors. These coils are rated at approximately 35KVs and represent the best of their type. The design of these coils reduces losses.

Electronic coils

Electronic coils are high-voltage coils for use with electronic ignition (no contact breaker points) systems and these coils can be built to give a very high output because they do not have to take into account the erosion of points. If this type of coil is fitted to a points type ignition system, the points will burn out very quickly.

HIGH TENSION WIRES

Fit new HT wires (leads) to any ignition system being used in a high-performance application. The quality of wires varies and original equipment or standard replacement wires are all going to be of the suppression type. For road use, suppression wires are universally used to stop interference to televisions and radios. All suppression high tension wires can be tested with a meter to check their resistance. Note that even new wires can be faulty so they should be checked too. If a wire has too much resistance (25K ohms and above) less power will be delivered to the sparkplug and the affected cylinder could produce less power as a consequence. 3-5K ohms of resistance in an HT wire is usually enough for good suppression. The less resistance in the wire the better, provided the wire is able to offer sufficient suppression and that's easy to check via a portable radio.

For competition, copper core wires (and derivatives) are still widely used and quite rightly so. Copper wires are excellent conductors of electricity and, essentially, cause no losses compared to suppression wires. Copper wires cause radio and television interference and are *not* suitable for use on the road.

SPARKPLUGS

The majority of sparkplugs which are readily available are suppression type sparkplugs. Sparkplugs that are not suppressed are available and can be used with suppression wire or with unsuppressed HT wires. There is no electrical continuity between HT wire connector and electrode with suppression type sparkplugs but there is with non-suppression sparkplugs. The majority of road-going engines have suppressed high tension wires and suppression sparkplugs fitted to them.

The concept of using copper HT wires with suppression plugs is well founded and is recommended for high-performance applications. The use of suppression wire, together with suppression sparkplugs, can be the cause of poor spark quality (even misfire) on high-performance engines which feature high compression (11:1 and up).

CHECKING SPARK QUALITY

With all new parts fitted and the engine up and running, certain checks are made to ensure that the ignition system is functioning correctly. Ultimately, what matters is how much power is delivered to the sparkplugs and that they are firing correctly. No performance engine will produce top power if the amount of spark is low. To fire a high compression engine (10:1 and up) the condition of the electrical system has to be perfect before any testing can be carried out; KV at the sparkplug must be known.

Wiring & connections

Fit brand new wires and new connectors wherever possible to the *whole* ignition system. Where wires are difficult to renew, such as in an existing wiring loom, if possible remove the old connectors, cut the wire back to expose a new clean copper core and put new connectors on. All terminals (ignition switch terminals, coil terminals, and so on) must be absolutely clean with no corrosion present. Breaks in wires, corroded and/or loose terminals can be the cause of a reduction in voltage or an intermittent electrical fault.

Alternator

All engines (road-going or competition) should be fitted with a good alternator. Road-going vehicles, for instance, have to have a good charging system

IGNITION SYSTEM

to continue running and the voltage measured across the battery terminals, while the engine is running and charging, should be 13.5-14 volts. An alternator which is putting out this amount of voltage is what is required, not one that is putting out just 12.1-12.2 volts at maximum rpm.

The practice of using a large capacity 12 volt battery instead of a generator for a competition engine is just not acceptable. The ignition system will have insufficient voltage and, although the engine will run, and run quite well, it will not run as well as when the voltage available is up to 13.5-14 or more across the battery terminals (a battery will hold 12.5 volts fully charged).

Fit a new or rebuilt alternator to the engine and check that the voltage output is as it should be; *do not* simply put a new alternator on to an engine and assume that the voltage output is correct. Check it and note the voltage across the battery with the engine running: check the output of the alternator whenever ignition troubles are experienced and see that the figures still match those originally recorded.

Alternators come in varying sizes and weights and, as lightness is a prerequisite for all racing engines, the smallest and lightest are usually sought for this purpose. Small, lightweight, high output alternators are fitted to many small-engined Japanese cars and some European cars (Citroen 2CV, for instance). All alternators can be slowed down by changing their pulleys for larger diameter ones so that the alternator is turned just sufficiently fast to supply a minimum of 13.5 volts across the battery at idle (however high or low that may be), and maintain a minimum of 13.5-14 volts at the rpm that the power band of the engine starts at.

Ignition switch

The ignition key/switch should always be considered a part of the ignition system because, if the ignition switch fails, it will not send 12 volts consistently to the primary side of the coil and this will show up as a very weak spark (low KV at the plug). Problems that often stem from the ignition switch (loose contacts) include intermittent faults, such as an engine misfire, hard starting (sometimes), no ignition (sometimes) and, eventually, no ignition at all. Vibration can cause the mechanism inside the switch to become loose and make poor electrical contact.

With a voltmeter check that there is 12 volts going to the primary ('+') side of the coil.

High tension current (points type distributor)

With a contact breaker points type ignition system a simple tool that can be used to check the integrity of the spark at the sparkplug is the Gunson Flashtest. This is an inexpensive tool made of plastic with a direct reading scale. The scale is proportional to the size of the gap and the gap, which the spark must jump, is increased or decreased by opening and closing the arms of the device.

Caution! Don't use the Flashtest gauge on an electronic ignition system. The module can be irreparably damaged. Use a Gunson Flashtest on points type ignitions only.

The KV available to the sparkplug is read directly off the Flashtest scale. If the KV is in the green part of the scale the ignition is okay, if the KV is in the white or red part of the scale there is something wrong.

To check the high tension (HT) sparkplug wires take the wire off a sparkplug (one at a time) and connect it to a terminal of the gauge, earth the other end of the gauge to the engine. Turn the engine over (sparkplugs out) and see what KV is present. What you will know if the gauge is in the green is that all parts of the ignition system before the end of the plug wire are in good order and the spark is sufficient. This, of course, does not mean that the sparkplugs themselves are okay. Note the KV reading of each plug lead.

High tension current (electronic type distributor)

With electronic ignition, the high tension has to be checked with a more sophisticated meter, such as the Snap-on Tools MT 2700 DIS/KV probe, for example, which can be used to test any ignition system. The meter uses an inductance pick-up which lightly slips over the ignition lead (or coil lead) and the dial on the meter is turned until the light stays on and is then turned the other way until the light starts to flicker. A reading is then taken. This meter does not give a true KV reading, but it does not matter that it doesn't. Essentially a reading of 2-3 KV means

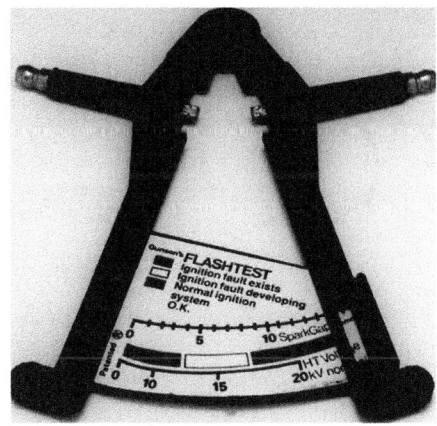

Gunson 'Flashtest' device is suitable for both points and electronic type ignition systems.

SPEEDPRO SERIES

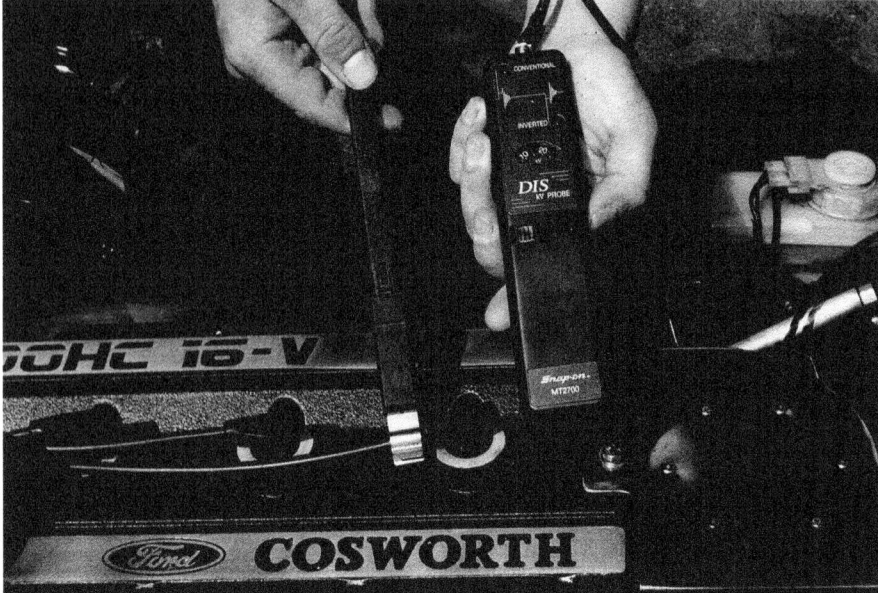

Gunson 'Flashtest' device (top) and Snap-on Tools inductance pick-up device (above) both being used to check voltage at the sparkplug end of an HT wire on a Sierra Cosworth-headed Pinto engine.

that there is a fault, while 8-12KV in the ignition wire means the system is working correctly. 20KV means that there is definitely something wrong. The coil HT wire will have readings of 10-16KV for a system that is working correctly, while 3KV means that something is wrong. 20KV-plus also means that something is wrong. Certainly the integrity of the spark can be checked with this equipment and faults easily diagnosed.

These types of meter will pick up the fact that the sparkplug is faulty. If the engine is missing, for instance, all of the sparkplug wires are checked for KV. A lead that gives a low reading should have the sparkplug changed and a further reading taken after the change. If the plug was at fault the reading will rise to match, approximately, that of the other leads. Alternatively, if a low reading is found and the sparkplug is difficult to get at, change the HT wire for a new one (resistance tested) and check the KV in the new lead. If the reading rises to match approximately the others the lead was at fault. If the reading stays the same the plug is at fault and will have to be replaced.

The output of the coil can be checked by slipping the inductance pick-up over the coil lead and taking a reading. The reading will frequently be the same as the plug lead reading or slightly above (2KV).

An alternative is to have the engine checked at an auto-electrical specialist which will use a sophisticated analyzer. This way all doubt will be removed as to the integrity of the spark. The cost for checking the engine will be reasonable and if there is anything wrong it will be quickly found.

One point about the electrical testing described so far is that it has all been carried out without the engine being under load conditions and an engine under load does not necessarily perform in the same way. The fact that electricity takes the path of least resistance always applies so what was regarded as a completely satisfactory performance in unloaded condition can suddenly become a quite unsatisfactory performance under load. When this happens all of the ignition system components, new or used, must be checked for faults.

IGNITION TIMING MARKS

These engines have degree markings on the front pulley/crankshaft damper

IGNITION SYSTEM

for the setting of static and idle speed ignition advance. There is no standard provision for the setting or testing of the total advance.

The crankshaft damper/pulley will have to be marked to include the increased idle speed advance degrees and the total advance degree markings. The amount of idle speed advance is not going to be more than 20 degrees while the total advance degrees is going to be 36 or 38 degrees before top dead centre.

Checking TDC markings

The timing of the engine is dependent on knowing where the true top dead center (TDC) point of the engine is. If there is error here, all other degree markings are inaccurate, but not necessarily useless if the degree of inaccuracy can be established. To avoid any confusion the TDC point *must* be checked. If the engine is being assembled this is a relatively easy task.

If the engine is assembled, and in the car, the procedure is a little more complicated but TDC can still be accurately found. The method involves the use of some resin core solder which is about 3mm(1/8in) in diameter. The solder is inserted through the sparkplug hole so that it will become wedged between the top of the piston crown and the cylinder head as the piston nears TDC. With the solder across the top of the piston the crankshaft is turned clockwise manually (plugs out) towards TDC. In fact, this is a dead stop method as the piston is not able to get to TDC because of the solder. The solder is soft and will not damage anything yet it will not crush unless force is applied to the crankshaft.

Once the piston contacts the solder, mark, with a white marking pen, the damper/front pulley adjacent to the fixed pointer or TDC line on the block/timing cover.

Now turn the engine anti-clockwise until the piston again contacts the solder, stopping any further rotation of the crankshaft. Mark, with a white marking pen, the damper/front pulley adjacent to the fixed pointer or TDC line on the block/timing cover. The true TDC line on the damper/front pulley has to be exactly in the middle of the two temporary white marks, otherwise it is incorrect. Check the marks again to be sure that no mistake has been made.

If the manufacturer's original timing marks are incorrect, the situation will have to be remedied. Note that a pointer can be bent slightly to reposition it, while a bolted on scale can be repositioned. Alternatively, the front pulley can be moved around the crankshaft by making an offset key, or the front pulley timing mark can be braised up and a new one made in the correct position.

Marking crankshaft damper/pulley

To ensure complete accuracy, the damper/front pulley should be removed from the engine (it can be marked with a reasonable degree of accuracy while still on the engine). The first thing that has to be done is to measure the diameter of the damper/front pulley.

Draw a circle the diameter of the damper on to a piece of paper using a compass. Use a protractor to mark on the circumference of the circle 38 advance degrees and then the idle advance degrees. This will give accurate dimensions which can be transferred to the damper/front pulley using engineer's dividers. This process is reasonably accurate (within 1 degree usually).

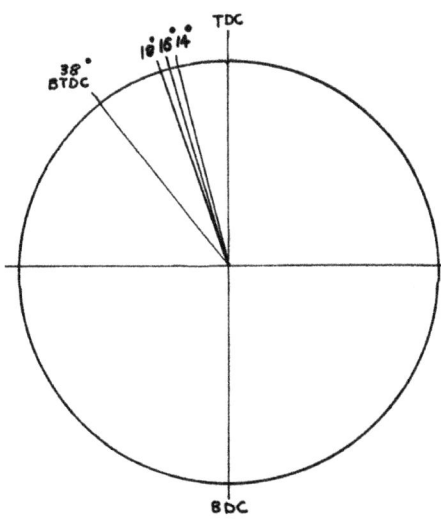

The sort of diagram you will make to show various degree markings for the crankshaft pulley. By placing the pulley face-down on the diagram, the marks can be accurately transferred to the pulley.

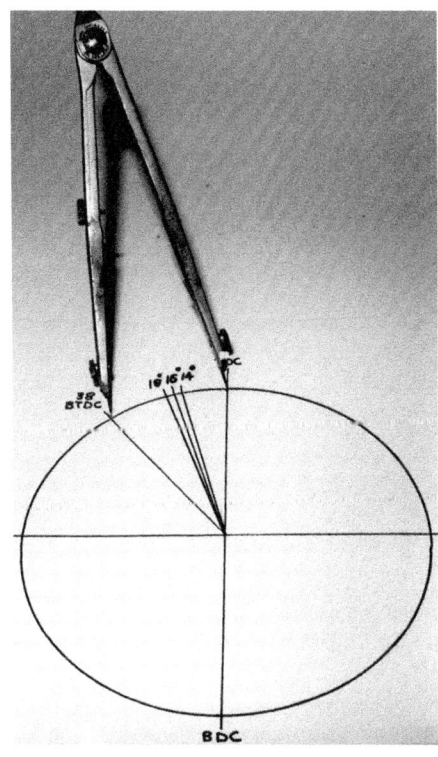

A pair of dividers being set to the correct distance as per the diagram.

SPEEDPRO SERIES

The dividers have been set to size on the diagram and are then used to transfer the degree marking distance to the crankshaft pulley.

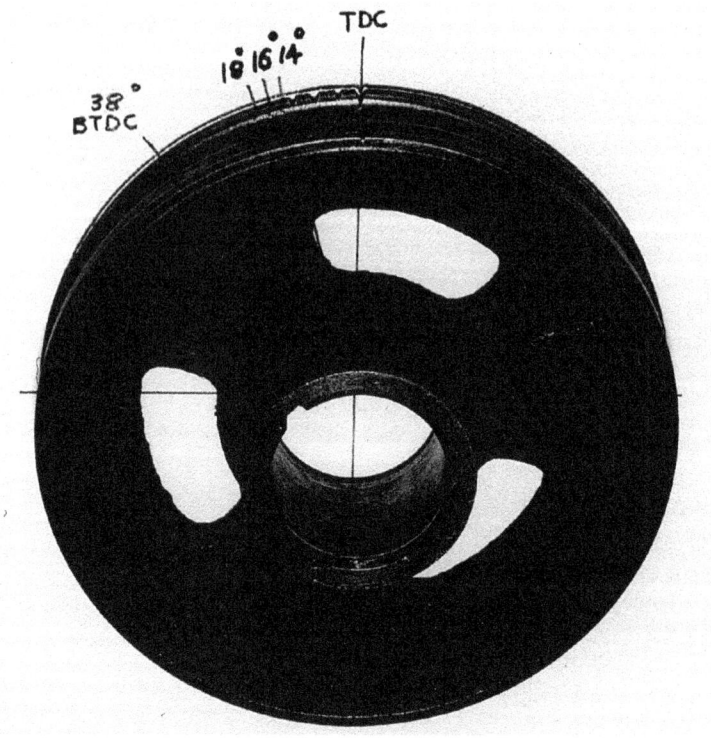

The crankshaft pulley placed face down on the timing diagram ready for the timing marks to be transferred.

Permanent advance degree marking

The crankshaft damper/pulley is removed and final markings are checked by placing the damper or front pulley face down on to the original piece of paper that the estimated advance marks were drawn on.

Remove the pulley/damper in order to make the final permanent timing marks – it's difficult to put deep and accurate marks into the pulley/damper when it's in place. When the pulley/damper is off, it can be placed face down on the timing diagram and the optimum timing marks transferred very accurately.

The final marks should be machined into the damper/pulley square to its front face and be at least 1.0mm (0.040in) deep. This is best done using a milling machine with the damper/pulley held in a machine vice and the grooves cut using a pointed 'D bit' cutter.

An alternative method of marking is to hold the damper in a vice and use a hacksaw to cut into the rim surface. The width of a standard hacksaw blade is ideal and the depth of the cut can be limited to 1.0mm (0.040in). Care must be taken to ensure that the hacksaw cut is square to the front face of the damper/pulley; you can use a small engineer's set square.

STATIC ADVANCE

The distributor (contact breaker points or electronic-type) will have to have an idle speed advance (vacuum advance temporarily disconnected) of between 12 and 20 degrees depending on the duration of the camshaft: the idle speed also varies depending on the camshaft.

For a standard (non-reprofiled) camshaft equipped engine, consider a 600 to 700rpm idle speed normal

IGNITION SYSTEM

and a minimum of 12 to 14 degrees of idle advance will be needed. Consider 18 to 20 degrees of advance to be the maximum ever needed for a competition engine fitted with a 310 to 320 degree duration camshaft and with an idle speed of between 1200 and 1400rpm.

The following should be helpful –

Camshaft	Idle advance degs
Std	12-14 degrees
285 degree	14-15 degrees
295 degree	15-17 degrees
305 degrees	17-18 degrees
320 degrees	18-20 degrees

TOTAL ADVANCE

All high-performance Pinto engines will need a full mechanical advance (centrifugal advance) of 36 or 38 degrees before top dead centre (BTDC). The full advance must be 'all in' at 3300-3500rpm. The 38 degree point has to be permanently marked on the crankshaft pulley and the ignition set so that this amount of mechanical advance is always present at anything above 3500rpm (vacuum advance, if fitted, is disconnected during setting of the ignition for full advance).

Note that the reason two figures are given here instead of one is that while most engines respond to 38 degrees, in some instances 36 degrees will prove to be just as good. Run the engine with both amounts of total advance and then use the amount of advance that 'suits' the particular engine (one usually better than the other).

VACUUM ADVANCE

The Bosch distributor has a vacuum advance mechanism and, for road use, it should be retained as it will give increased economy (up to 15 per cent). It was popular at one time to disable vacuum advance systems in high-performance applications. This was done because the vacuum advance mechanism is connected to the distributor baseplate and, on some distributors, allowed too much fluctuation of the advance; to prevent this fluctuation, the whole advance mechanism was removed and the baseplate brazed up to prevent any movement at all. The Bosch distributor is of excellent quality and does not suffer from vacuum advance related problems when installed and working as it should.

With a twin sidedraught carburettor set up (or any other with non-standard manifolds), the vacuum advance system can be successfully operated by tapping into one of the individual inlet tracts and running the usual small diameter pipe between it and the distributor.

The vacuum advance system can be expected to increase the mechanical spark advance by approximately 10 to 15 degrees when the engine is operating under vacuum (such as when cruising). When acceleration is demanded from the engine, there is no vacuum and the mechanical advance is all that's operating (38 degrees BTDC if the engine is revving over 3500rpm or so).

IGNITION TIMING SETTING AND CHECKING

With the crankshaft pulley accurately marked at 12, 14, 16, 18, 20 and 38 degrees BTDC, it's a simple operation to check and then reset, if necessary, the ignition timing using a stroboscopic timing light ('strobe'). The strobe is an essential piece of equipment if the timing is going to be set accurately. On any competition engine check the ignition timing before the event starts, just to be sure that the timing is still correct.

With a marked pulley and using a strobe, the timing can be checked to see exactly what number of degrees of advance are present at idle speed and at various higher rpm (not more than 3700rpm) by checking the rpm on the rev counter (tachometer); the speed at which full advance is reached can also be checked.

REV-LIMITERS

It is important that engine rpm be limited to avoid connecting rod breakage and valve float.

Bosch make governor rotors the lowest of which is set at 6200rpm. This type of limiting device cuts the high tension current when the set engine speed is reached, killing the engine until the speed drops. The spring in the rotor arm can be 'tweaked' to increase the tension and, as a consequence, the rpm that the engine cuts out at will be increased.

The alternative to a governor rotor is to fit one of the electronic limiting devices that are readily available. The fitting of both is not a silly idea.

IGNITION SYSTEM SUMMARY

Consider 12 degrees of idle advance to be the minium required for use on any standard (not reprofiled) camshaft Pinto engine. Such an engine will idle at 500-600rpm.

Consider 20 degrees of idle advance to be the maximum for any modified Pinto engine. Engines with this much idle advance will have an idle speed in excess of 1200rpm and up to 1500rpm and, as such, the centrifugal advance will be in operation (just); as a consequence the idle advance is more than the static advance.

SPEEDPRO SERIES

The total mechanical advance setting for any standard or modified Pinto engine is 38 degrees.

The vacuum advance will advance timing by up to an extra 15 degrees BTDC under cruise conditions.

The vacuum advance will advance timing by only 3-5 degrees (depending on camshaft and idle speed) at idle.

The vacuum advance must be disconnected when the idle ignition timing and total ignition advance is being set with a strobe light.

Maximum rpm should be limited by fitting a governor rotor arm or an electronic limiting device or both.

Chapter 13
Carburettors

Twin twin choke downdraught Weber carburettors. Hard to find and offering no real advantage over sidedraughts.

The ideal high-performance Pinto carburation system uses either twin 40 or 45 Dellortos or Webers. There are aftermarket electronic fuel-injection systems available for these engines but, for cost reasons, the sidedraught carburettor is what is fitted in the vast majority of high-performance applications.

Well cast and machined aluminium inlet manifolds are available which bolt on to the cylinder head with only minor port matching being required. Aftermarket inlet manifolds are usually cast to suit 45mm carburettors, which means that when 40s are fitted there is some mismatch but no detrimental affect.

There have been a few other induction systems used on modified Pintos over the years, such as twin twin-choke downdraught carburettors but, these days, these are rare and expensive to buy if you can find them and offer no advantage in performance terms.

For further, in-depth, information on Weber DCOE and Dellorto DHLA carburettors another Veloce Publishing SpeedPro Series book is available – How To Build & Power Tune Weber & Dellorto DCOE & DHLA Carburetors, 2nd edition, by Des Hammill.

Within this chapter you'll find basic carburettor settings which will suit all Pinto engines, whether in standard form or modified.

All 1600, 1800 and 2000 Pinto engines can use twin 40mm carburettors with 34mm diameter chokes, however, 1600 engines are best suited by 40mm carburettors. The fitting of 45mm carburettors to 1600 engines is not recommended, as they'll prove slightly less efficient than 40mm carbs. The fitting of 36mm chokes in

SPEEDPRO SERIES

Twin Weber DCOEs on RS2000 Pinto engine. (Peter Phillips' car)

45mm carburettors will prove to be too much for this size of engine, unless the engine is highly modified and being run frequently to constant high rpm. For all-round use, 40mm carburettors will prove better than 45s.

Modified 1800 engines can use 36mm chokes in 45mm carburettors but, once again, such an engine will need to be highly modified to benefit from the bigger carbs. The fitting of 40mm carburettors fitted with 34mm chokes will generally prove suitable for almost all applications (up to 145bhp).

The following basic carburettor specifications have proved successful and most are based on real conversions. Consider them as a starting point for your own application -

• **Jetting for standard 1600/1800/2000 engines (40mm Weber DCOEs)**
34mm chokes.
135 main jets.
F11 emulsion tubes.
190 air correctors.
35 accelerator pump jets.
40 F9 idle jets.
4.5 auxiliary venturis.
Float level shut off height - 7.5mm.
Float height at full droop - 15.0mm.
Idle screws turned out 7/8 of a turn.

Example engine had 12 degrees of spark advance at 600rpm and 38 degrees total mechanical advance.

The distributor is a Bosch electronic (part designation G FU 4) and the vacuum advance is connected and operating.

• **Jetting for a modified 1600cc engine (40mm Weber DCOEs)**
34mm chokes.
140 main jets.
190 air correctors.
F16 emulsion tubes.
40 pump jets
40 F9 idle jets.
4.5 auxiliary venturis
Float level shut off height - 7.5mm.
Float height at full droop - 15mm.

• **Jetting for standard 1600/1800/2000 engines (40mm Dellorto DHLAs)**
34mm chokes.
140 main jets.
7772.10 emulsion tubes.
180 air correctors.
7850.1 idle jet holder.
40 idle jets.
7848.1 auxiliary venturis.
40 pump jets.
Float level shut off height - 15mm.
Float height at full droop - 25mm.

Example engine had 12 degrees of advance at an idle speed of 600rpm and 38 degrees of total mechanical advance. The distributor was a Bosch points type (part designation J FU 4) with vacuum advance.

• **Jetting for a modified 1800 engine (45 Weber DCOEs)**
36mm chokes.
140 main jets.
170 air correctors.
F16 emulsion tubes.
40 accelerator pump jets.
45F11 idle jets.
4.5 auxiliary venturis.
Float level shut off height - 7.5mm.
Float height at full droop - 15mm.
Idle screws each 1 full turn out.

CARBURETTORS

Example engine had 18 degrees of idle advance and 38 degrees of total advance. No vacuum advance was fitted.

- **Jetting for a modified 1800 engine (45 Dellorto DHLAs)**

36mm chokes
145 main jets.
180 air correctors.
7772.6 emulsion tubes.
40 accelerator pump jets.
8011.1 auxiliary venturis.
50 idle jets.
7850.1 idle jet holder.
Float level shut off height - 15mm.
Float height at full droop - 25mm.

Example engine had 18 degrees of idle advance and 38 degrees of total advance. No vacuum advance was fitted.

- **Jetting for a modified 2000 engine (45mm Dellorto DHLAs)**

38mm choke size.
150 main jets.
7772.6 emulsion tubes.
190 air correctors
40 or 45 accelerator pump jets.
60 idle jets.
7850.9 idle jet holder.
8011.1 auxiliary venturis.
Float level shut off - 15mm.
Droop setting - 25mm.

Example engine had 18 degrees of advance at an idle speed of 1200rpm and a total mechanical advance of 38 degrees. There was no vacuum advance fitted to the engine and fuel economy was down because of this.

- **Jetting for a modified 2000 engine (45mm WeberDCOEs)**

38mm chokes.
145 main jets.
180 air correctors.
F16 emulsion tubes.

40 accelerator pumps.
45 F2 idle jets
4.5 auxiliary venturis.
Float level shut height - 7.5mm.
Float height at full droop - 15mm.
Idle screws turned out one full turn.
50 accelerator pump inlet.

Example engine had 18 degrees of advance at an idle speed of 1200rpm and 38 degrees of total mechanical advance. No vacuum advance was fitted (removed from the distributor) and the economy was down because of it.

THROTTLE ACTION

Check that a wide open throttle is actually being achieved in all carburettor chokes when the throttle pedal is fully depressed. Lack of full throttle opening is a common problem that frequently robs engines of full performance. Check the linkage setting frequently just to be sure that the carburettor butterflies are fully open at full throttle.

CARBURETTOR SUMMARY

40mm sidedraughts (with 34mm chokes) are the smallest to consider for any standard Pinto engine (1600, 1800 or 2000cc).

The use of 34mm chokes in 40mm sidedraughts for standard 1800 and 2000 engines is well founded and will give top performance. The trend is to think that 45mm side draughts are necessary for these larger engines, but this is not so.

A modified 1600cc engine will also use 34mm chokes in 40mm sidedraughts with appropriate jetting. The fitting of 45mm sidedraughts with larger chokes (36s or 38s) to 1600 engines is virtually never successful unless the engine is turning very high rpm all of the time.

A modified 1800cc engine will be able to use 35mm or 36mm chokes in 45mm sidedraughts but, virtually, never chokes bigger than 36mm.

A modified 2000 or 2100cc engine will almost always use 38mm chokes in 45mm sidedraughts. The fitting of 48mm sidedraughts is virtually never necessary, or desirable.

Consider the full range of chokes suitable for use in standard or modified Pinto engines to be from 34mm through to 38mm.

Note that choke sizes go up in 1mm increments in sidedraught carburettors so the amount of airflow can be optimised relatively easily through testing. Use the smallest choke that gives the best overall performance.

Larger chokes, such as 40s, when fitted to modified 2000/2100 engines will almost always cause the engine to perform worse at under 7800-8000rpm than it would with 38mm chokes. The trend is to think that bigger is better when it comes to chokes, but this is seldom correct.

INLET MANIFOLDS

Manifolds for twin sidedraught carburettors usually have the front two runners angled to allow access to the distributor, although straight runners would be better. There's a wide choice of aftermarket manifolds but be aware that some are designed for maximum convenience rather than efficiency. When choosing your manifold bear in mind that the ideal manifold will have runners as straight as possible, 75-100mm/3-4in long and with internal dimensions that taper down from carburettor butterfly size to port size.

Fit the inlet manifold (without carburettors) to the cylinder head and check to see that there is no mismatch between the manifold and inlet ports of the head. The inlet manifold runner diameters will be between 36 and 38mm in the area closest to the head. If any mismatch is found (treat the manifold runners as a continuation of the cylinder head ports) remove material from the manifold as appropriate.

Make sure that the smallest diameter of every manifold runner is no less than the choke size being used in the carburettors.

AIR FILTERS

All engines should be fitted with a good quality air filtration system. It only takes one small stone to be ingested into the engine to cause major damage.

Air filters suitable for twin sidedraught carburettors are available from many companies such as Piper X, Ramair, K&N and IR to name a few.

Good air filters prevent engine damage and excessive wear in valve seats, cylinders and rings.

Modern air filters cause virtually no restriction to airflow.

RAM PIPES

Ram pipes (tubes) are recommended as they improve airflow. They come in a variety of shapes and sizes but your choice will normally be determined by the space available within the air filter. As a general rule, fit the longest ram tubes that can be accommodated.

FUEL SUPPLY

Weber and Dellorto carburettors require good fuel volume, but not high pressure.

Consider a supply pressure of 1.5-2.5psi to be ideal and at least 1.5psi should be maintained under full throttle conditions. If the pressure drops below 1.5psi, the fuel level in the float bowls will drop, creating a lean mixture.

Pressure can be measured by fitting the take-off of a fuel pressure gauge between the two carburettor unions (or banjos) and the gauge itself to the dashboard so that pressure can be monitored in various situations. The gauge and extra fuel line can be removed once satisfactory fuel pressure has been established.

Chapter 14
Sierra Cosworth & Cosworth-headed Pinto engines

INTRODUCTION

The twin camshaft, four valve per cylinder Sierra Cosworth engine is a derivative of the standard Sierra overhead camshaft engine and proves the point that, with some re-engineering, the 'Pinto' engine is excellent. Around 39,000 Cosworth engines were built and, while no longer produced by Ford, these engines are still much sought-after for use in all sorts of applications where good solid power is required. These engines were only ever factory fitted in turbocharged form.

These engines are now disappearing quite quickly from the scene, with few cars ending up in scrapyards. The cylinder heads and blocks, or complete engines, when they do come up for sale, are selling for good money. These engines certainly have their following.

This chapter deals specifically with conversions which will be naturally aspirated (use carburettors) although many of the modifications would also be appropriate to turbocharged and fuel-injected engines. The requirements for conversion to natural aspiration are covered extensively. When fully modified in naturally aspirated form, this engine produces levels of power and reliability matched by few other mass-production engines

Sierra Cosworth head and cam cover.

of similar design. Although this engine is not light, it's very strong and can rev to high rpm (9000) and still be considered virtually unbreakable.

The genuine Cosworth block has thick walls (but no thicker than a late model 165, 185 or 205 Pinto

SPEEDPRO SERIES

Sierra Cosworth head fitted with carburettors and mated with a Pinto cylinder block.

Raised top, 12 to 1 piston for a naturally aspirated Cosworth engine.

block): later Cosworth blocks are numbered 200 and the early ones 205. The crankshaft is forged and its journals are induction hardened, the connecting rods are very strong and the pistons are forged. The Cosworth crankshaft has nine flywheel bolts and the flywheel carries an 279mm/11in clutch. The bellhousing is aluminium.

Despite hard use (and often poor maintenance) the majority of Sierra Cosworth engines do not fail, so when they are found in breaker's yards it's usually because the car in which they were fitted was involved in a crash and written-off: the cars are much quicker than most drivers realise (until it's too late!). Many 'recycled' Sierra Cosworth engines are converted to natural aspiration and then used to power high-performance and competition cars of all descriptions.

There are a number of Sierra Cosworth cylinder heads around in the second-hand market and these can be successfully fitted to non-Cosworth Pinto blocks to very good effect, though rpm will be limited by the strength of the bottom end. These heads are a direct replacement for the Pinto item but will require all the head modifications detailed in this chapter, together with some modifications to the cam drive system. With this conversion the previous chapters dealing with block preparation should be followed.

This chapter describes the procedure for modifying a genuine Sierra Cosworth engine to accept natural aspiration by way of alternative pistons and cylinder head modifications. Modification of the Sierra Cosworth cylinder head by porting, camshafts, camshaft followers, exhaust systems, carburation and ignition are all covered extensively.

The standard unported Sierra Cosworth cylinder head on a naturally aspirated engine will allow the engine to produce 150 to 170bhp, but it will never match the power output given by a well ported 'cammed-up' cylinder head on an otherwise identical engine. The difference between the two engines (modified head versus unmodified head) is night and day and as a consequence only the modified cylinder headed engine is covered in the conversion procedure.

The camshafts recommended for these cylinder heads have 'mechanical' profiles, which means replacing the standard hydraulic followers with solid items. Revs will depend entirely on the strength of the bottom end. A genuine Cosworth engine converted to natural aspiration will turn 9000rpm reliably, while non-Cosworth Pinto blocks will be subject to the rpm limitations detailed in the Pinto section of this book.

Summary of modifications

For natural aspiration usage the inlet and exhaust ports need to be enlarged.

Expect the cost of modifying the cylinder head correctly to be considerable because there is a lot of work involved (a lot of metal to be carefully removed).

CR needs to be increased from the standard 8 to 1 ideally to a minimum of 10 to 1. However, in some instances, genuine engines have

SIERRA COSWORTH & COSWORTH-HEADED PINTO ENGINES

simply had the block planed to bring the compression up to about 9 to 1 and have still gone very well with this amount of compression.

The standard inlet and exhaust valves are well sized and for almost all applications do not need to be larger.

The port sizes recommended in this book are suitable for road and competition use. Ports can be opened up with reasonable safety (not breaking through into waterways, etc) because the head casting is substantial.

Genuine Sierra Cosworth engines should ideally be fitted with forged raised top pistons to increase the compression from the standard 8 to 1. These pistons are 'drop in fit' items and the valve reliefs will all have been correctly positioned by the piston manufacturer.

COMPRESSION RATIO (CR) – PINTO & COSWORTH BLOCKS

There are several options which will increase the compression ratio (CR). How much to increase CR depends on the application and the octane rating of the fuel to be used. Too much compression is worse than too little because the engine will suffer pre-ignition (pinking/detonation) and engine damage is likely.

These engines ideally need 10 to 1 through to 11 to 1 compression to perform really well. Consider 11:1 to be the ideal figure to have if the inlet valve is closed at between 75 to 80 degrees ABDC. If 11:1 compression is built into an engine and there is no pinking under full acceleration (meaning the fuel being used is of excellent quality), the engine performance will invariably be very good.

The compression ratio *must* be matched to the octane rating and

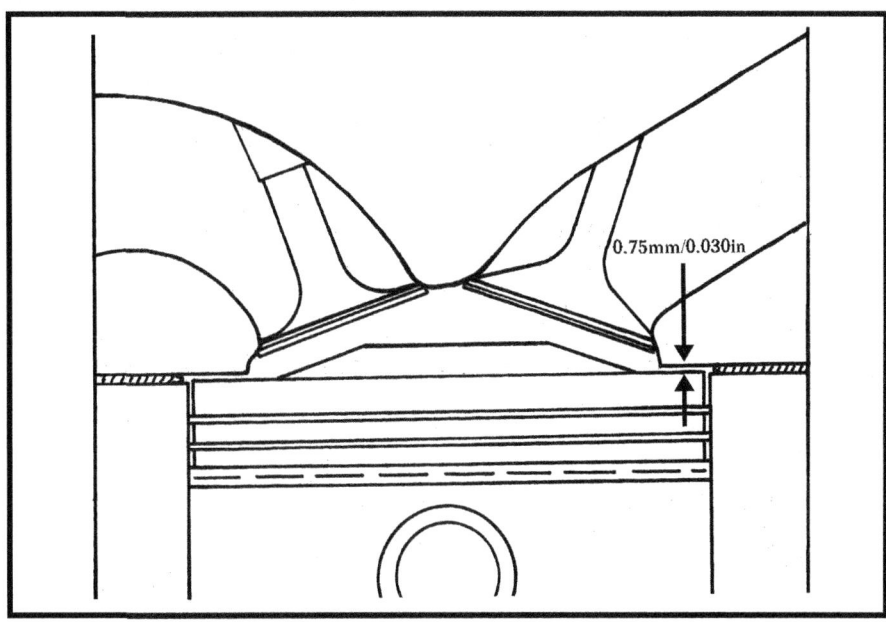

Minimum clearance between piston crown and cylinder head.

quality of the fuel to be used. By way of further explanation, 'all-in' spark advance of 32 degrees BTDC is generally the best setting for maximum power but if, with your chosen fuel, the engine pinks under full acceleration it will be necessary for the CR to be reduced. There is no point in not being able to accelerate the engine using full throttle because the engine is pinking continuously, but *do not* retard the ignition to stop such pinking because performance will be lost. The engine is over compressed and the only solution to the problem is to reduce the compression ratio.

A standard 2000cc Pinto engine short block assembly will have approximately 9.95 to 1 compression once four valve reliefs have been cut into the crowns. Four valve reliefs have to be machined into the tops of the pistons to prevent piston to valve contact if camshafts with longer duration than standard are going to be fitted. The depth of all four valve reliefs being a minimum of 3.5mm/0.140in.

The use and preparation of a standard Pinto short block assembly, while being a reasonably inexpensive option, does result in an engine limited in rpm through having the standard connecting rods and pistons fitted. There is, however, no doubt whatsoever that, up to the rpm limit, such an engine will go very well.

Genuine Sierra Cosworth engines that are going to be used in all out competition are invariably fitted with 12 to 1 Cosworth or Accralite pistons (or similar) and are capable of turning 9000rpm reliably. The valve reliefs are deep and correctly positioned for a standard deck height block. Consider 12 to 1 compression to be the absolute maximum to ever use (and only then with fuel that will allow this amount of compression).

These same pistons can be fitted into any 2000cc block and will result in approximately 10 to 1 through 10.5 to 1 compression (depending on the type of piston being used). The connecting rod's piston pin tunnel

SPEEDPRO SERIES

Auxiliary driveshaft (left) nose has been turned down to take a genuine Sierra Cosworth sprocket (right).

sprocket has to be changed to a Sierra Cosworth-type to suit the Cosworth drivebelt (which must be used) and the Pinto auxiliary shaft will have to be turned down to allow the fitting of a Sierra Cosworth-type auxiliary drive sprocket. While the standard Cosworth drive belt is almost always adequate, stronger belts are available. Burton Power, for example, offers a high strength belt which has carbon-fibre in it. These belts are slightly wider than the standard ones but still fit the standard drive sprockets.

CYLINDER HEAD PORTING

Caution! The porting work (to the dimensions which follow) must by carried out by an experienced professional as, in some areas, wall bore will have to be honed out so that there is a piston pin clearance of 0.010mm/0.0004in so that the forged piston pin retention method can be used. The crown of the piston will, of course, be 1.5mm down the bore more than usual, since the Cosworth connecting rods are 1.5mm longer. The block can be planed to increase the compression (1.5mm/0.060in with complete safety). If 11.8 to 1 or 12 to 1 pistons are fitted into a standard Pinto short assembly, the fact that the piston crowns are down the bore means that a sometimes useful reduction in compression is the result. The planing of the block by 1.0mm/0.040inch might well prove to be a more appropriate amount.

Inlet port with right-hand tract opened out.

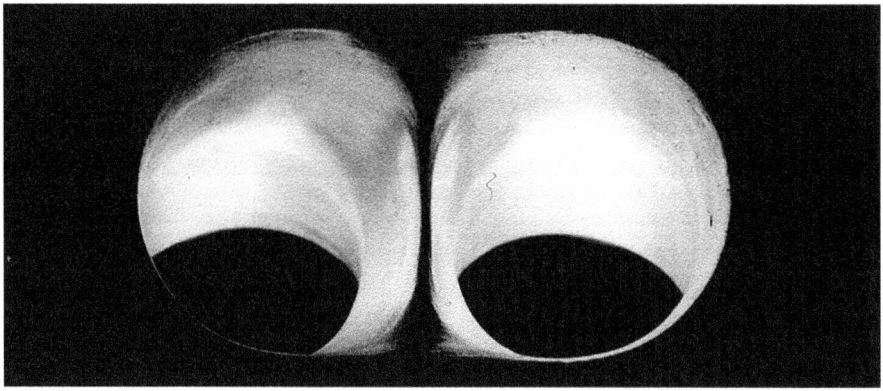

Completed inlet port viewed from manifold end.

COSWORTH HEAD/ PINTO BLOCK CAMSHAFT DRIVE MODIFICATIONS

If a Sierra Cosworth cylinder head is fitted to a Pinto block the crankshaft

SIERRA COSWORTH & COSWORTH-HEADED PINTO ENGINES

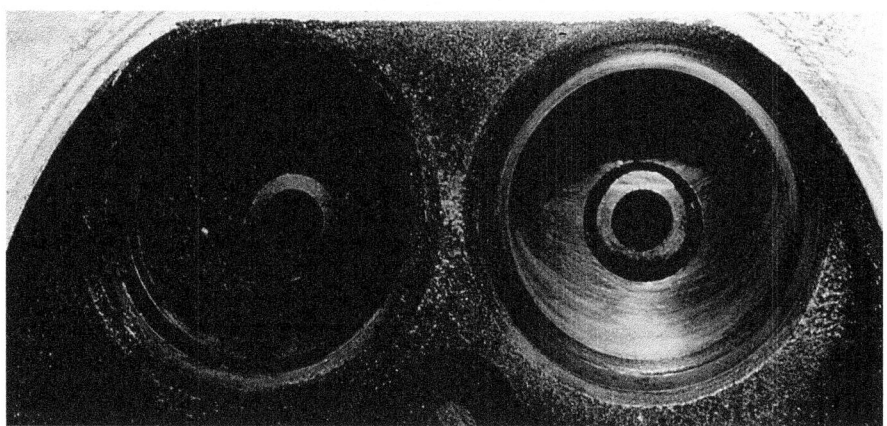

Another view of the completed inlet port viewed from manifold end.

Completed inlet port (right) compared with standard.

Inlet ports

The stock inlet port entry at the side of the head is oval in shape and measures 49.5mm by 25.0mm. In the modification process, this is taken out to the size of the gasket (which should be used as a template) aperture which is 50.5mm by 26.5mm wide. The top edge of the inlet port opening can be contoured as shown in the photograph in an effort to keep the port roof as high as possible. This contour runs into the inlet manifold for 32mm/1.25in.

As the inlet port progresses inward to the point of bifurcation, the stock diameter is approximately 22.8mm/0.900in. Increase the overall diameter of this section to 26.8-27.3mm/1.050-1.075in which is a considerable increase in the cross-sectional area of each port. In broad terms the opening up of the individual inlet ports to these sizes is the main factor in gaining better volumetric efficiency.

The inlet port then turns into the valve seat area and, directly adjacent to the valve seat, is an area which is oval in shape and which measures 26.0mm/1.025in by 27.3mm/1.075in. This area is opened out so that it becomes round and is 30.0mm/1.180in in diameter. The valve seat is not modified in any way except for a reduction in width to 1.2mm/0.048in (the seat's outer diameter should be the same as the valve's – standard is 35mm/1.377in).

To check whether or not each individual port has been made large enough, the following simple test can be carried out. Make a gauge with a thin section 26.0mm diameter washer trapped between two nuts on a length of threaded stud. The gauge is passed down each inlet port from the inlet manifold end: keep the washer as

thickness gets thin and a mistake could cause breakthrough. The inlet valve guides do not need to be removed during porting work but there's no doubt that it's easier with the exhaust guides removed. That said, if left in place, the exhaust port guides can be worked around.

A considerable amount of aluminium has to be removed from the inlet and exhaust ports which takes time. There's plenty of material in these cylinder heads and, with *careful* porting work, very large ports are possible and there will still be sufficient wall thickness for reliability.

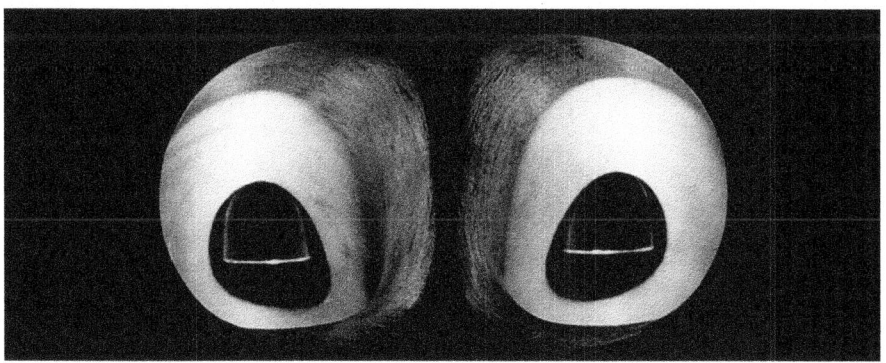

The completed exhaust port viewed from the manifold end.

SPEEDPRO SERIES

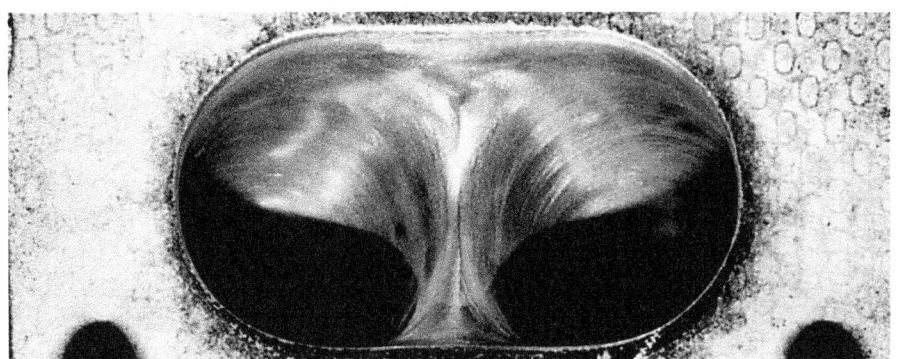

Another view of the completed exhaust port viewed from the manifold end.

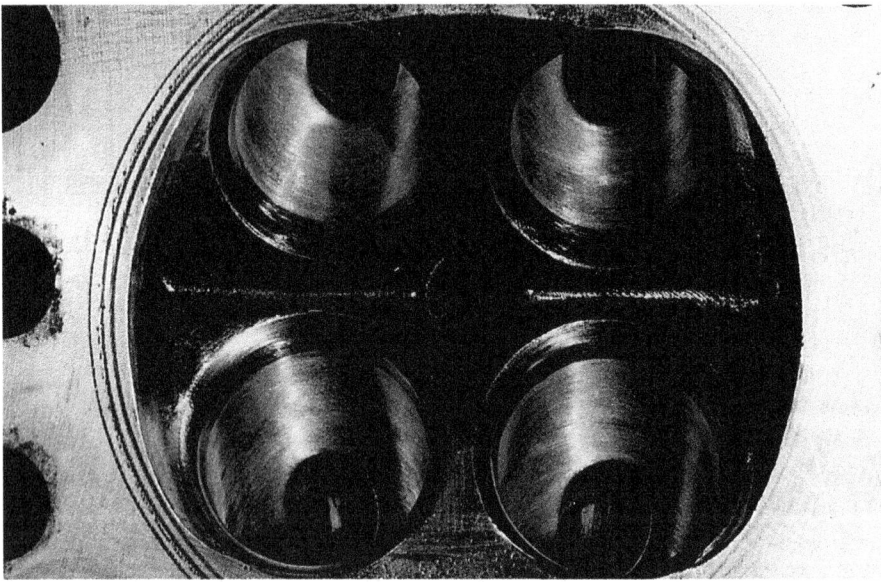

Completed exhaust and inlet ports viewed from the combustion chamber.

square to the port as possible. The fit of the gauge in the port must be what is termed a 'rattling good fit' (hence the size of the gauge versus the finished size of the port): the gauge must pass unhindered through the port and past the valve guide to the valve seat.

Exhaust ports

The original port exit shape at the side of the cylinder head is oval and the standard dimensions are, approximately, 48mm/1.965in by 24.5mm/0.970in. The port is not rough in finish at all. The port exit is opened out to the standard exhaust gasket aperture size (take material *only* from the top and sides of the exhaust port) – you can use the gasket as a template. The width of the port exit is increased to 51mm/2.010in and the height to 27mm/1.060in.

The exhaust ports are 24.0mm/0.945in in diameter at the narrowest part of the valve seat. The valve seat insert is opened out to 26.7mm/1.050in in diameter.

The valve seat width is reduced to 1.5mm/0.060in whilst retaining an outer diameter to match the valve's - standard 31mm/1.220in. The narrowest part of the exhaust port is just at the 90 degree point of the turn in the port and the diameter here is 19.5mm/0.765in. Open the port to a minimum diameter of 24mm/0.950in (taking material equally from everywhere and flare out to the enlarged port exit. The floor of the exhaust port should be flattened off as much as possible. The port's wall thickness in the 'corners' can get quite thin in the porting process but a reasonable estimate of the port wall thickness can be made by measuring (through the water holes) with a vernier caliper the depth from the gasket surface of the head. There is not a limitless amount of aluminium wall thickness (4 to 5mm nominally) in these cylinder heads, but certainly 2.0mm can be taken off everywhere except the exit base. The completed port is far removed in shape and size from the standard port.

CAMSHAFTS
Valve lift limitations

The standard camshafts have a minimum amount of duration, have 'hydraulic' profiles and feature

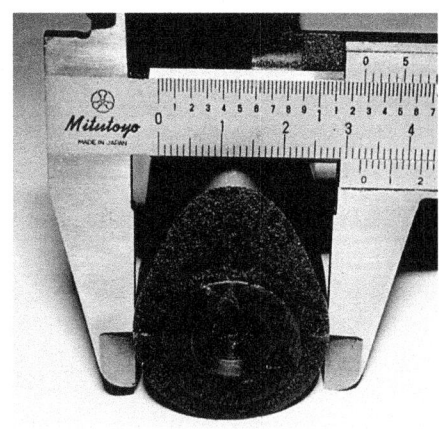

Measuring lobe base circle.

SIERRA COSWORTH & COSWORTH-HEADED PINTO ENGINES

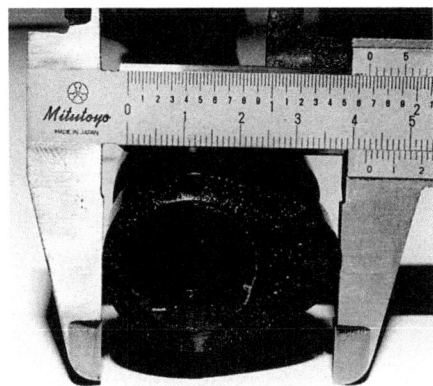

Heel to toe measurement of lobe.

8.5mm/0.333in of valve lift. Inlets open 8 degrees BTDC and close 52 degrees ABDC. The exhausts open 52 degrees BBDC and close 8 degrees ATDC (note that this sequence of events (in degrees) is the most common in the description of camshaft event timing so all alternative event timings in this text are set out in the same order). This is a very mild camshaft specification which is suitable for the standard turbocharged engine but is *not* suitable for a modified naturally aspirated engine which is required to GO.

The standard camshafts have plenty of material for reprofiling so any alternative camshaft profile can be ground on to the original camshafts. The actual camshaft lobe is of very generous proportions, having a base circle diameter of approximately 38mm/1.495in. The measurement from the 'heel' of the camshaft to the 'toe' is approximately 46.5mm/1.830in.

Caution! The camshaft recommendations that follow do *not* take into account piston to valve clearances. Suffice it to say that *no* long duration, high lift camshaft should ever be fitted to one of these engines and the engine run until it is known that sufficient piston to valve clearance exists (minimum of 2.25mm/0.090in at TDC on either valve).

Note that although Piper Cams and Kent Cams are specifically mentioned in order to assist readers to get parts, on a worldwide basis, of the right specification first time, many other companies produce excellent camshafts and related parts which are direct equivalents.

There is often some variation between camshaft grinders' figures for the same camshaft profile. This is because of slightly different starting points for the measuring process and should not be taken to mean that the camshaft regrinder who lists more degrees of duration is actually selling you a camshaft which is effectively different and, by implication, better!

Camshafts with up to (8.94mm/0.352in) lift can be used with standard valve springs if the engine is partially dismantled to remove the valve spring platforms, and the backs of the platforms ground down to 0.6mm/0.024in thickness from their original 1.0mm/0.040in.

Any camshaft which has a valve lift of 9.0mm/0.355in, or more, will need to be used with alternative valve springs. The standard Sierra Cosworth valve springs coil bind at around 22.0mm/0.866in and, with the standard camshafts fitted, the compressed height of the standard valve spring is 23.25mm/0.916in.

Standard Sierra Cosworth valve springs limit the valve lift to that of the standard camshaft (slightly more with valve seat work). However, mechanical camshaft profiles are readily available which do offer more duration with the standard amount of valve lift and can be used successfully.

High-performance camshafts (low lift)

The standard Sierra Cosworth valve springs can be retained if, for absolute reliability, lift is limited to a maximum of 8.89mm/0.350in. This means that camshaft profiles such as the BD3, L2 or BP300, for example, can all be used with the standard Sierra Cosworth valve springs. All these camshafts have no more than 8.89mm/0.350in valve lift, but L2 and BP300 will need the backs of the valve spring plates to be ground down to give extra coil clearance to avoid binding.

Both Kent and Piper, for instance, offer identical camshafts by profile type name but they list the duration degrees differently which can be confusing. The BD3 has 290 degrees duration when listed by Kent Cams and 284 when listed by Piper: both have valve lift of approximately 8.58mm/0.339in. Kent Cams BD3 inlet opens 35, closes 75; exhaust opens 75, closes 35. Piper Cams BD3 inlet opens 32, closes 72; exhaust opens 72, closes 32. The full lift timing point is the same for both BD3 camshafts at 110 degrees.

The L2 profile has 302 degrees duration and valve lift of 8.73mm/0.344in when listed by Kent Cams and 294 degrees duration and valve lift of 8.89mm/0.350in when listed by Piper. The recommended full lift timing position is the same for both pairs of camshafts at 102 degrees. Kent Cams L2 inlet opens 49, closes 73; exhaust opens 73, closes 49. Piper Cams L2 inlet opens 45, closes 69; exhaust opens 69, closes 45.

Piper's BP300 camshaft has 304 degrees duration and valve lift of 8.89mm/0.350in. These camshafts are timed for full lift at 104 degrees. Piper Cams BP300 inlet opens 48, closes 76; exhaust opens 76, closes 48.

To improve the low rpm (2000 to 7500rpm) performance of an engine fitted with the BP300 camshafts they can be phased for later inlet closing

SPEEDPRO SERIES

and later exhaust opening. Note that the degrees of overlap remain the same but the timing cycle as a whole has been moved around. This is one way of improving the overall performance of the engine. Set the camshafts at the exact number of degrees for the inlet closing and the exhaust opening and forget about the overlap degrees. Piper Cams BP300 (with altered phasing) inlet opens 45, closes 79; exhaust opens 71, closes 51.

All of these camshafts have 'mechanical' profiles so standard hydraulic followers will have to be replaced by 'mechanical' or 'solid' followers (buckets) but engine dismantling of any consequence is not necessary, so the conversion to mechanical camshafts to gain duration and different lift rates is relatively easy.

Camshafts with considerable 'overlap' (the time in degrees that the inlet valve and the exhaust valve are open together) can cause a rough idle.

All of the foregoing camshafts are of the low lift category but, used in conjunction with a well modified head, will give excellent power to 8500rpm (around 290 degrees) or 9000 (around 300 degrees).

Once the camshaft to be used is chosen, order it by name such as 'BD3' or 'L2.'

High performance camshafts (high lift)

Caution! Any camshaft with lift of more than standard (8.89mm/0.350in) will necessitate the use of non-standard valve springs to avoid coil bind.

These camshaft profiles are not specific to the Sierra Cosworth engine but are from other engines such as the Lotus Cortina (L1 and L4), Ford BDA engine (BD4) or the Cosworth DFV (DA10).

290 degree plus duration camshafts
An example is the RS500 GPA3 turbo camshafts which have 292 degrees of duration with valve lift of 10.67mm/0.420in. These camshafts are quite suitable for naturally-aspirated engines turning up to 8500rpm.

GPA3 - inlet opens 40, closes 72; exhaust opens 72, closes 40.

300 degree plus duration camshafts
For use with high revs (up to 9000rpm) are the L1 mechanical grinds, and similar, which have 302 degrees of duration as listed by Piper Cams/306 by Kent Cams. They have valve lifts of 10.06mm/0.396in for the inlets and exhausts.

Note that Kent Cams' phasing of these camshafts is definitely correct at 47-79-71-55 for the Sierra Cosworth engine (tried, tested and proven!). **Caution!** Make sure there is plenty of piston to valve clearance on the exhaust valves.

310 degree plus duration camshafts
These are competition/full race camshafts. To take full advantage of the longest duration camshafts the rpm that the engine is turned to has to be as high as 9000rpm on a continuous basis, and the engine has to be strong enough (competition prepared) to stand this sort of treatment reliably.

Full race camshafts all have valve lifts of at least 10.0mm/0.395in and up to a maximum of 11.00mm/0.430in. If the engine will be turned to 9000rpm at every gearchange, then the 312 to 316 degree camshafts will prove correct.

Dual valve springs and rev limiters must be fitted to these high rpm engines to avoid over-speeding. These engines will all have raised top pistons fitted, a minimum of 11:1 compression and plenty of piston to valve clearance.

There are a few profiles that fall into the maximum duration category and two examples are the DA10 and BD4. These camshafts have advertised durations of 312 and 316 respectively. DA10 inlet opens 54, closes 78; exhaust opens 78, closes 54. BD4 inlet opens 56, closes 80; exhaust opens 80, closes 56.

These two camshafts have more than enough duration for any naturally aspirated Sierra Cosworth engine being turned up to 9000rpm. The opening point of the exhaust valve is quite important and so is the closing point of the inlet valve. The earlier the inlet valve is closed (conducive to good cylinder filling) the better, while the later the exhaust valve is opened (conducive to good cylinder clearing) the better. Consider exhaust opening 80 degrees before BDC to be the

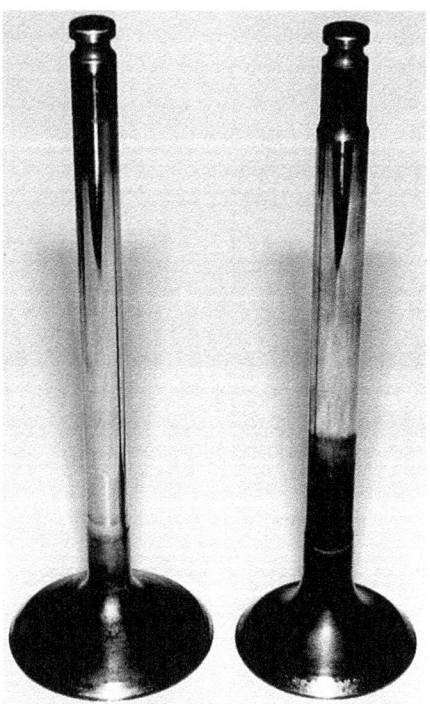

Standard valves.

SIERRA COSWORTH & COSWORTH-HEADED PINTO ENGINES

earliest the exhaust valves should ever be opened and consider 80 degrees after BDC the latest that the inlet valves should ever be closed.

An alternative option is to fit an exhaust camshaft which has 306 degrees of duration (the L1 for example) whilst retaining one of the foregoing inlet camshafts. Such a set-up improves the torque of the engine throughout the rpm range with a slight power loss at high rpm.

(eg: DA10 inlet profile and an L1 exhaust profile - inlet opens 54, closes 78; exhaust opens, 71, closes 55.

VALVES
Standard valves

The standard inlet valve is 35.0mm in diameter and the standard exhaust valve is 31.0mm in diameter. The standard valve seat insert will safely allow the use of a 36.0mm diameter inlet valve and the standard exhaust valve seat insert will safely allow the use of a 32.0mm exhaust valve. The valve stem diameters are not the same for inlets and exhausts: inlets 7.0mm and exhausts 8.0mm. The use of standard sized valves is recommended for any engine being turned up to 9000rpm. These valves have what are termed standard length stems. The inlet valve weighs 62 grams and the exhaust weighs 56 grams. The exhaust valves are sodium filled. The standard valves are of excellent quality and suitable for the vast majority of applications.

It's advisable, and recommended, that all of the valves be replaced with brand new ones when rebuilding an engine for high-performance. If you are intending to use used valves, the valve stem diameters *must* be checked and, if wear is present, the valves written off. Each valve is measured at the top (near the groove) and, with this original size known, the full length of the stem is checked. Expect to find wear (if present) near the valve head or about 25mm/1 inch away from the actual head. Any wear from new or original size is *unacceptable*.

Holbay make one piece stainless exhaust valves with 31.5mm/1.250inch diameter heads, in either standard stem configuration or long stem configuration, with 9/32in/7.14mm valve stem diameters. The original guides need to be K-Lined down to suit these sizes They also make one piece stainless inlet valves with 37mm/1.400in diameter inlet heads, in either standard stem configuration or long stem configuration, with 9/32in/7.14mm diameter valve stems. They make collets/keepers turned from high tensile solid round bar. Mechanical followers to suit are also available from Holbay.

Long stem, bigger diameter valves

Slightly larger than standard diameter valves (non-Ford) are available. They have 35.5mm diameter (0.5mm/0.020in larger than standard) inlet valve heads and 31.5mm valve heads (0.5mm/0.020in larger than standard) for the exhaust valves. These valves also have long valve stems which can only be used in conjunction with mechanical profile camshafts and the type of lightweight cam follower that does not have a long central pillar. Both inlet and exhaust valves have 7.14mm/0.281in valve stems which means a valve guide change for the exhausts and a K-Line insert in the inlet guide.

The largest valve head diameters the standard seats will take is 36mm inlet and 32mm exhaust. The stems

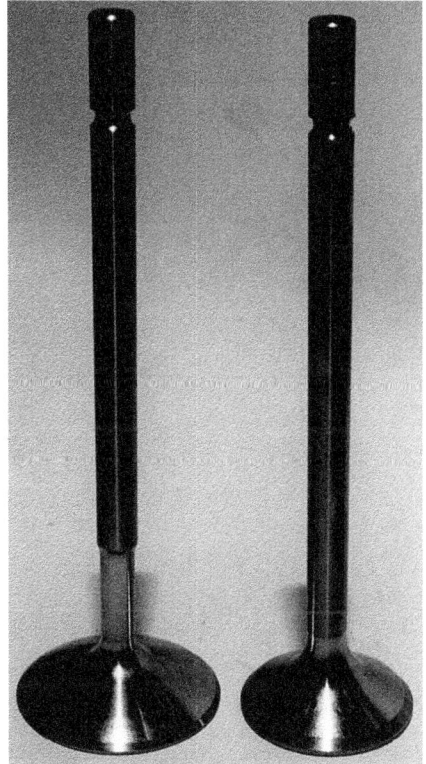

Long stem valves.

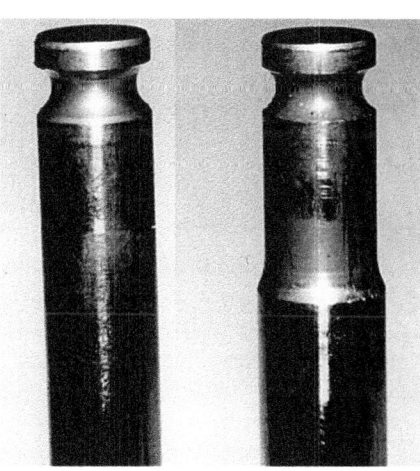

Valve stem keeper groove detail.

Valve stem seals.

SPEEDPRO SERIES

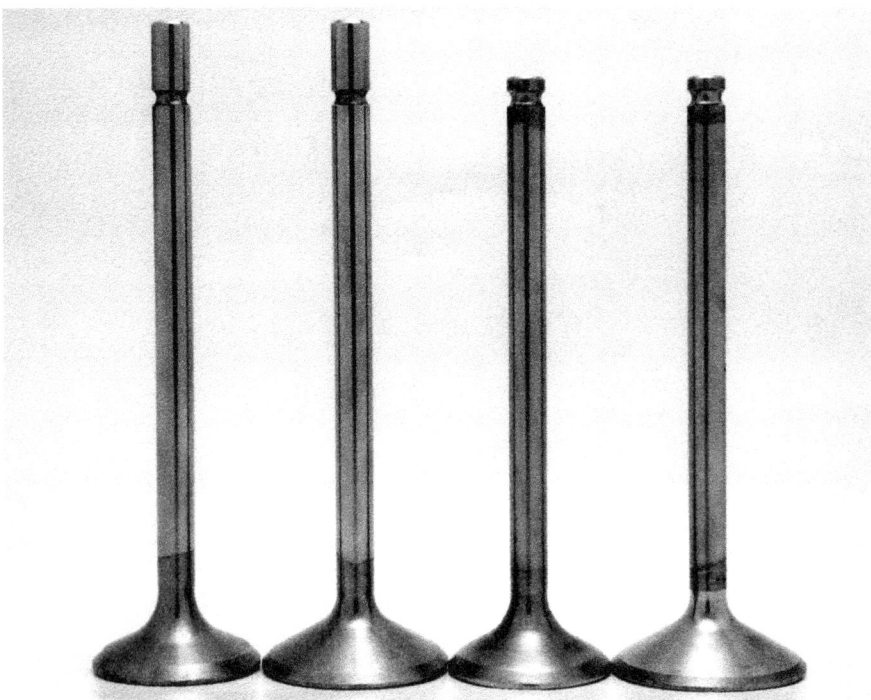

Holbay inlet and exhaust valves for Sierra Cosworth cylinder heads. Long stem on the left and standard stem on the right.

of these valves are long enough to go right up to near the top of the camshaft follower and valve clearance is effected by changing to a very small shim. This system is very light overall and is recommended for all very high rpm engines (9000rpm). The valve stem diameters are 7.0mm/0.275in which will mean that the exhaust valve guides will have to be changed. Valve clearance is obtained by changing or grinding down the shim that fits between the top of the valve stem and the underside of the cam follower.

Valve keeper grooves

While both standard valves have different stem diameters, they have the same diameter keeper grooves (the stems of the exhaust valves are reduced at the top). The grooves in all valves standard or otherwise are the same. The standard valve keepers and valve retainers are of excellent quality and are suitable for all applications.

Valve stem seals

There are two standard sizes, one for the 7mm inlets and one for the 8mm exhausts. The inlet seals are red and the exhaust ones are green. The fitting of inlet valve type seals to the exhaust guides is required once the valve stem diameter of the exhaust valves has been changed to 7.14mm/0.281in.

Valves - summary

The retention of the standard valves, valve retainers and keepers is recommended for all applications up to 9000rpm.

Standard valve head diameters are 35mm for the inlets and 31mm for the exhausts.

Solid cam followers are available for use with standard length valves.

Long stem valves are available with larger head diameters: 35.5-38mm for inlet valves and 31.5-35.5mm for exhaust valves.

Long stem valves are used in conjunction with a different type of solid follower.

VALVE SPRINGS
Standard valve springs

The standard engine uses single springs which have a free height of approximately 40mm/1.575in and a 30.0mm/1.18in outside diameter. The fitted height of the springs is approximately 31.5mm/1.245in and the seated spring pressure is approximately 65 pounds. With the standard 8.5mm/0.335in valve lift the pressure is 145 pounds when the valve is fully opened ('over the nose' of the camshaft lobe). The valve springs are compressed to 23.0mm/0.903in at full standard lift and are coil bound at 21.5mm/0.850in. This means that the standard valve springs are only safe for a *maximum* valve lift of 8.5mm-8.7mm/0.333in-0.340in.

Top - standard spring platform (left) and upside-down ground platform (right). Above - standard platform and shim which can be used to balance spring pressures or to replace platform.

SIERRA COSWORTH & COSWORTH-HEADED PINTO ENGINES

The valve lift with which the standard valve springs can be used can be increased a little by removing the standard spring platforms and grinding their backs down. The standard thickness of these spring platforms in 1.0mm/0.040in and they can be ground down (using a surface grinder) until they are 0.4mm/0.015in thick. This corresponds to an increase in total useable valve lift to 9.1-9.3mm/0.358-0.365in before coil bind. The 'over the nose' valve spring pressure will be unaltered and seated valve spring pressure will be slightly less, but the reduction is not cause for concern for use up to 8300rpm. Alternatively, the spring platforms can be removed from the cylinder head and replaced by 0.25mm/0.010in hardened steel valve spring shims which are usually a direct replacement (depending on the inside diameter of the central hole).

Note that the original spring base has three small 'dimples' which prevent it from sitting flat and, when they are ground off, an extra 0.2mm/0.008in is effectively gained in the fitted spring height. These dimples indent the surface of the cylinder head,

Four alternative types of valve spring. Left to right - standard Sierra Cosworth; standard Sierra Cosworth with an inner spring added; Iskendarian outer only; Iskendarian dual spring.

but the spring platforms never actually sit flat because of the deformation in the spring platform itself.

Up to 8300rpm is feasible with standard valve springs (depending on the camshaft profile and actual 'over the nose' spring pressure).

Valve springs for high lift camshafts

The fitting of high lift camshafts and the speed that the engine is going to be turned to both need to be taken into account when choosing springs.

There are two categories of application: 1) high lift (more than 8.5mm) camshafts in engines that are not going to be revved above 8000rpm; 2) high lift (more than 8.5mm) camshafts in engines that are going to be used for competition and revved to high rpm (9000rpm).

Singles (up to 8000rpm)

There are single valve springs available that will fit into the cylinder head without any modification (Kent Cams VS30 for example). They have 32.5mm fitted height (which isn't that important) and do not coil bind until approximately 18.0mm/0.708in (which is very important): the Sierra Cosworth valve springs are coil bind at approximately 21.52mm/0.850in.

This type of valve spring allows the use of camshafts with lifts up to 11.0mm/0.430in but the spring pressures are similar to those of the standard valve springs, so rpm *must* be limited to 8000.

Duals (up to 9000rpm)

Dual valve springs have to be used for high rpm operation. Examples of such

Standard valve retainer (left) and machined down version (right).

Standard Cosworth valve spring.

Iskenderian dual valve spring.

springs are Piper Cams VDSCOS, Kent Cams VS34 or Iskenderian (Cross Flow Ford) dual valve springs all of which have a fitted height of 34.0mm/1.338in (not the 31.5mm of the standard engine).

These valve springs are not direct replacements. There are two ways in which the 34mm fitted height can be accommodated. The first is the use of different valve retainers (with a higher spring seating) such as Kent Cams Titanium VRT02, for example, and the second is that the spring bases of the cylinder head have to be machined deeper (in which case all of the standard components are used).

In short, these three valve springs will actually fit into the engine but the 'over the nose' pressure will be too high (250 to 260 pounds) when used in conjunction with an 11.0mm/0430in high lift camshaft. Consider that 200 pounds of over the nose pressure as more than adequate for 9500rpm operation: the maximum ever likely to be required. Any engine that is used with such high spring poundage will exhibit signs of wear in the camshaft caps (shiny rub marks) which are a clear indication of too much valve spring pressure. Many engines seem to get fitted with springs with this type of poundage, but it's wrong.

It is pretty much accepted that when any dual valve spring is fitted the installed height is going to be non-standard and, while alternative retainers do sometimes solve this problem, many cylinder heads end up having their spring seats deepened by a machining operation.

Deepening the spot facing of the spring base by using a suitable trepanning tool on a milling machine is one way of increasing the fitted height of dual springs. The valve seat recesses in the cylinder head are remachined 2.0-2.5mm/0.080-0.100in deeper and the standard valve retainers are re-used. These valve springs have a coil bind height of approximately 19.5mm/0.770in so, with the standard fitted height of 31.5mm, this would allow for an absolute maximum travel to coil bind of 12mm/0.475in and a useable valve lift of 10.5mm. After machining the cylinder head, this distance is increased to 33.5-34.0mm/1.320-1.335in and the absolute maximum travel to coil bind is 14.0mm/0.550in.

With dual valve springs installed at the recommended 34.0mm fitted height the seated valve spring pressure will be approximately 65 pounds and the 'over the nose' pressure approximately 200 pounds with a 11.0mm/0.430in lift camshaft.

An often overlooked possibility is that, with the cylinder head valve spring seats machined deeper, the standard Sierra Cosworth single valve springs can be used provided a suitable inner valve spring is used in conjunction to give the necessary increase in pressure. With the standard spring now having a fitted height of 34.0mm (instead of 31.5mm) there will be a total of 12.5mm travel before coil bind and a useful camshaft lift potential of 10.7mm. A suitable inner valve spring will have a free-length of 35-37mm, be left-hand wound and have an outside diameter of 21mm. With the standard valve spring retainer such a spring will be compressed to 31mm at the installed height and have 15 pounds of seated pressure. Compressed a further 10mm (as it would be to suit a 10.0mm lift camshaft) the pressure must be 60 pounds which adds up to an 'over the nose' pressure of 195 pounds (200 pounds with a 10.5mm lift camshaft) at full lift. Note that the inner valve springs of the Piper Cams, Kent Cams and the Iskenderian dual valve springs recommended for use in these engines all fit inside the standard Cosworth valve spring and each will provide the required tension.

Iskenderian dual valve springs have a free length of 42.5mm/1.675in for the outer and 35.5mm/1.400in for the inner valve spring. They are not a perfect fit with the standard valve retainer so the retainer needs to be machined slightly to suit the inside diameter of the outer valve spring. The outside diameter is 28.0mm/1.100in and the inside diameter15.2mm/0.600in. Coil bind height is 18.5mm/0.730in.

Dual Iskenderian valve springs are an excellent choice because they can, one way or another, provide valve spring pressure to suit most applications and are readily available everywhere. As an example of the flexibility of these springs, if the outer valve spring only is fitted into the standard cylinder head (31.5mm/1.243in fitted height), the seated pressure will be approximately 70 pounds (5 pounds more than standard) and at 8.5mm/0.333in compression, 150 pounds of 'over the nose' pressure will be applied which is slightly more than standard. At 10.0mm/0.395in of valve lift with these outer valve springs there is 160 pounds of 'over the nose' pressure and 3.0mm more travel before coil bind. The Iskenderian outer valve spring coil binds at 18.5mm.

It is quite acceptable to use just the outer valve springs and this option allows the use of a camshaft with up to 11.0mm/0.432in of lift with moderate valve spring pressure (maximum 8500rpm).

If dual valve springs are fitted to an engine which has standard 8.5mm lift camshafts, there will be approximately 205 pounds of 'over

SIERRA COSWORTH & COSWORTH-HEADED PINTO ENGINES

the nose' pressure (approximately 60 pounds more than standard) which is acceptable even though it's the maximum figure recommended in the interests of reliability. The seated valve spring pressure will be approximately 100 pounds.

Note that if dual valve springs are going to be used with low lift camshafts (8.5mm) the standard spring platform should be removed and replaced by 0.25mm/0.010in thick flat shims of 31.75mm/1.250in in diameter and with a 13.46mm/0.530in diameter hole in the centre. This increases the fitted height of the springs by approximately 1.0mm/0.039in and reduces the seated spring tension to 95 pounds and the 'over the nose' pressure to about 195 pounds. No engine that has 8.5mm of valve lift is going to be revved to more than 9000rpm and this 'over the nose' pressure will safely allow this.

Measuring valve spring pressure

Irrespective of the make of the valve springs, all can be measured for poundage at certain heights. The standard fitted height is 31.5mm/1.24in

Inverted solid follower for Sierra cosworth. Top, follower and adjustment shim; above, shim in fitted position.

and the lift of camshafts ranges from 8.5mm/0.334in (standard) to about 11.0mm/0.433in. Once the camshaft choice has been made the lift will be known and so will the rpm requirement of the engine.

Escort diesel follower with adjustment shim in place.

Most engine machine shops (engine reconditioners) have a valve spring testing machine which, although not often used, is still usually found decorating a bench top.

As an example, if the spring's fitted height will be standard at 31.5mm/1.24in and the camshaft lift will be 10.0mm/0.393in, then the valve spring is placed on the tester and compressed to 21.5mm and a pressure reading recorded to see if it is in accord with the poundage/rpm requirements of the engine. Also test the coil bind height and the poundage just before coil bind. No valve spring should be fitted to an engine *without* testing to see how much pressure it has at the fitted height and, more

Face of solid lifter for Sierra Cosworth engines with standard length valves.

Left, standard hydraulic lifter; centre, Escort diesel follower; right, solid follower for Sierra Cosworth. Note how much shorter the hydraulic lifter is.

SPEEDPRO SERIES

importantly, at the 'over the nose' height.

The valve springs are also being tested to see whether or not they all give enough pressure and whether or not all of the valve springs have the same pressure at the same height. If several valve springs are found to have slightly less pressure than the others the pressure can be restored by using shims to pack the springs up – this causes the pressure to rise according to the thickness of the shim. There are limits to the amount of shimming that can be done because the coil bind height is reducing by the thickness of each shim you add. In fact, it is unlikely that valve springs are out by very much (usually 10 pounds maximum).

Valve springs – summary

Maximum permissible 'over the nose' pressure is 200 pounds. Maximum engine speed is 9500rpm.

Standard valve spring seated pressure is 65 pounds and 'over the nose' is 145 pounds.

Iskenderian dual valve spring specifications (Piper and Kent dual valve springs are rated similarly) -
• Outer valve springs only (31.5mm installed height) offer seated pressure of 70 pounds, an 'over the nose' valve spring pressure with 8.5mm lift of 150 pounds and 165 pounds with 11.0mm lift. By removal of the standard spring platform and replacement by 0.25mm/0.010in shims the spring pressure is reduced.
• Dual valve springs (32.5mm installed height) have a seated pressure of 95 pounds and an 'over the nose' valve spring pressure with 8.5mm lift of 190 pounds.
• Dual valve springs (34.0mm installed height) have a seated pressure of 70 pounds, an 'over the nose' valve spring

Bare Sierra Cosworth head.

pressure with 8.5mm lift of 175 pounds and 200 pounds with 10.0mm lift.
• Standard Cosworth valve springs with a fitted height of 34mm and fitted with a suitable inner, will give a seat pressure of 70 pounds and an 'over the nose' pressure at 10.0mm lift of 195 pounds.

CAMSHAFT FOLLOWERS

There are two types of mechanical profile camshaft followers made specifically for this engine. The first of these is designed to be used with the standard valves and, more to the point, the original stem length of these valves. The second type is designed

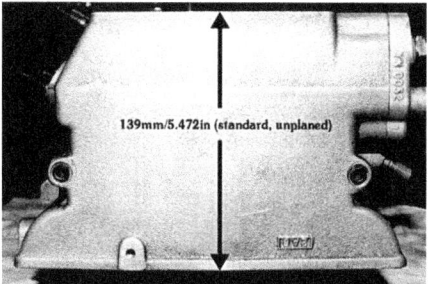

Thickness of a standard, unplaned Sierra Cosworth head is 139mm/5.472in.

to be used with the non-standard long stem valves. The difference between the two camshaft followers lies in the fact that one has a long centre pillar on the underside of the camshaft follower.

Both types of camshaft follower are very light and strong; however, the type used with long stem valves and the valve clearance adjustment shim are slightly lighter in combination and can rightly be regarded as the ideal parts for a competition engine. The camshaft followers used with the standard valves are, however, suitable for all competition applications and can be regarded as just as good up to 9000rpm. In the majority of instances, the standard valves will be retained and the appropriate camshaft follower used in conjunction with them.

There is another mechanical camshaft follower that can be used with standard length valves (*not* long stem) and it is the type fitted to the 1800cc Escort diesel engine. Similar in design to the previously mentioned followers, this has the addition of a top fitting valve clearance adjustment

SIERRA COSWORTH & COSWORTH-HEADED PINTO ENGINES

shim to facilitate valve clearance adjustments without taking the camshaft out (if fitted to the Sierra Cosworth cylinder head the camshafts still have to be taken out to adjust the valve clearances).

The reason these Escort cam followers are often chosen is cheapness (they can come from a scrapped car). Provided used camshaft followers are not worn (no longer absolutely parallel) through high mileage, they can be a good choice but they are heavier (around 90 grams, depending on shim thickness) than the two specifically-designed camshaft followers. They are quite acceptable for use with milder duration camshafts. These followers are also available from Ford.

The valve clearance adjustment shim of the Escort follower is simply lifted out of the camshaft follower and directly changed for a thicker or thinner shim to make a valve clearance alteration. The disadvantage is that the follower gets heavier as the shim gets thicker and there's also a limit to how thick the standard shims go. These camshaft followers are only suitable for use with Ford's standard range of shims if the camshaft's base circle is not less than 35.0mm/1.375in. Ford dealers stock shims that range in size from 3.0mm to 4.75mm in 0.05mm increments.

Escort followers can be used with any mechanical camshaft, irrespective of how much the base circle has been ground down (meaning below 35.0mm) provided custom-made shims are machined up but this is not recommended.

CYLINDER HEAD REBUILD

In the first instance, the cylinder head should be visually checked for cracks and for welded or other repairs and then checked thoroughly using crack testing equipment: there is no point spending time on a cylinder head that is not sound – 'bad parts are worse than no parts.' The cylinder head should also be pressure tested just to be sure there are no water leaks. Lastly, the cylinder head thickness should be measured to see if it has been planed.

The camshafts should be checked for free rotation (all cam followers removed) to ensure that the camshaft bearing bores are indeed straight and not out axially. Camshaft bores that are out axially indicate a warped cylinder head, usually caused by overheating. At worst, the camshaft bearing bores may have to be corrected by line boring.

Check all the head's stud holes for stripped threads. These are repairable by using Helicoils. Check the valve guides for wear and expect them to be well worn unless recently repaired or replaced.

Check the valve stems for wear especially near the head of the valve: the replacement of all valves with brand new ones is recommended. If a used valve shows measurable wear it *must* be replaced.

Valve guides

The valve guides are a high wear area and will normally require replacement or restoration of those guides that are not in tolerance. Consider any inlet valve guide that has more than 0.05mm/0.002in and any exhaust valve guide that has more than 0.055mm/0.0025in clearance as being too worn for further use. Guide bores wear oval so wear is actually quite difficult to measure. Unless the guides fitted to the head exhibit as-new characteristics restore them by a reconditioning process. New guides will have ideal clearances of 0.025mm/0.001in for the inlets and 0.0375mm/0.0015in for the exhausts.

Do not remove any valve guides unless they are loose in the cylinder head or damaged in some way which necessitates removal and replacement. K-Line guide inserts will restore original clearances in the existing guides and can be fitted by many engine reconditioners.

New replacement valve guides, as supplied by Ford, come semi-finished which means that the internal bore of the guide is not to finished size. Considering the cost of the guide, the cost of removing the guide, the risk of possible damage to the bore in the cylinder head and the cost of fitting the guide and final reaming to size, it's just not worth doing, especially when the K-Line guide inserts are just so much easier to install, last longer than original guides and can be replaced many times for little cost. Holbay make replacement valve guides.

Replacement valve guides

New valve guides are not easy to machine and the hole that is factory drilled in the 'centre' of the guide is not necessarily concentric with the outside diameter. This means that to open the hole out to suit the valve, the guide cannot simply be set-up in a lathe and drilled and then reamed out to size. The hole in the centre of the guide has to be machined true by boring (using a sturdy but obviously small boring bar). This method of sizing the bore of the guide is time-consuming and tedious, but *essential* if the bore is to end up concentric with the outside diameter of the guide.

Failure to get the bore true with the outside diameter of the guide will (if such a guide is installed in the head) result in the valve and the valve seat not being concentric by a wide

SPEEDPRO SERIES

and unacceptable margin. The valve seats then have to be reground to suit the new guide which can mean that a considerable amount of the valve seat has to be removed to true it up. Avoid this situation by ensuring that any replacement valve guide is accurately sized. The guides *must* be accurately machined inside (diameter wise) and the inside bore *must* be concentric with the outside diameter.

Removing valve guides

Before removing guides, be sure it is *really* necessary. Note that guides are sometimes removed to carry out major porting work.

Any valve guide that has to be removed from the cylinder head *must not* be pressed out. This means setting the cylinder head up on a turret-type milling machine (Bridgeport, or similar), boring (from the combustion chamber side of the guide) to 12.00mm-12.20mm/0.470in-0.478in inside diameter (the outside diameter of the guide is 13mm); the wall thickness of the guide is then only 0.40mm-0.50mm/0.015in-0.020in. The guide is bored to a depth of 34mm/1.340in from the front edge; this then leaves a good-sized internal step with which to drive the guide out. By using this method, the guides are virtually being 'collapsed' (being devoid of any circumferential strength, no longer hold any real interference fit in the cylinder head). The cylinder head should be heated to about 220F/105C using a propane torch before drifting the weakened guides out.

It is frequently recommended that, to facilitate guide removal, the cylinder head be heated to 220F/105C and the old intact guides driven out using a pilot-shanked drift punch. The punch has a pilot section that fits neatly into the guide for its full length. This system

Sierra Cosworth having the old guides machined to accept K-Line inserts.

of valve guide removal is all very well, but there will nearly always be some damage (galling of the bore) to the aluminium cylinder head, which is not good at all. After the guides have been removed and replaced a few times in this manner, the size of the bore in the cylinder head will be lost. The removal of guides using this method is not recommended for Cosworth heads: the risk of damage and the cost of a replacement cylinder head (or the remachining the valve guide bores) make it a less than ideal proposition.

Caution! Internal damage to the surface of the guide bores is to be avoided at all costs because, once the interference fit of the guide to cylinder head is lost, the cylinder head cannot

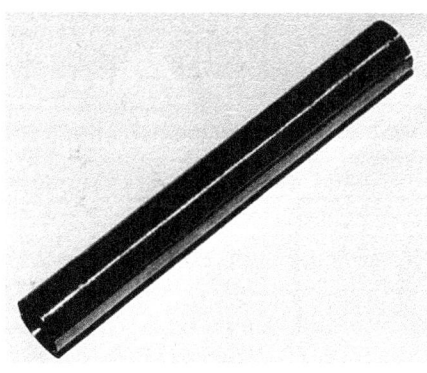

K-Line valve guide insert.

be safely used. If a guide comes loose in the cylinder head during service, a considerable amount of time, effort and expense is required to remedy the situation. This situation is avoidable by finding out the true status of the engine components when the engine is apart, and repairing or replacing *all* the items necessary to make the engine totally serviceable *before* it is assembled – don't cut corners!

To restore a damaged cylined head valve guide bore the original bore has to be located as accurately as possible using a special machine for valve guide reworking (Kwikway Head Shop, or similar) or a milling machine (Bridgeport, or similar) and be re-bored, and perhaps reamed, to suit a custom-made oversized valve guide – an expensive process and unnecessary if care is taken in the first place).

The valve guide has an interference fit in the cylinder head of 0.04mm-0.042mm/0.001in-0.0012in which is not really all that much and all the more reason to be as careful as possible when handling the cylinder head.

Fitting valve guides

Valve guides are driven into the cylinder head after the head has been heated to 220F/105C and the guides placed in a freezer (which will reduce the size of the guide by only 0.0025mm/0.0001in).

Before the valve guides can be fitted, a special installing tool must be made. This tool locates into the bore of the guide but the contact surface *must not* be the top edge of the guide (usual method), instead it's the shoulder further down the guide. This means that the original drift used to remove the guides can be re-used to install the guides with the aid of a specially made ferrule that is loose fitting around the top section of the guide and longer than the top section of the guide.

The guide will always have to be driven into place, but the blows should be light. If the guide virtually falls into the bore of the head, the chances are there is insufficient interference fit.

Before drifting the guide in, use clean engine oil to lightly lubricate the guide bore in the cylinder head. This will ensure that the guide does not 'pick up' in the aluminium of the cylinder head.

Restoring valve guides

The bore of the valve guide can be restored to as new, or better, by the use of very hard-wearing K-Line valve guide inserts. The original guide does not need to be removed from the cylinder head which precludes many potential problems. Find an engine reconditioner which has the necessary installation equipment. K-Line guides are available in 8.0mm/0.314in and 7.0mm/0.28in bore sizes which covers the range of these engines.

These inserts are replaceable as often as necessary and, once the internal bores of the original guides have been appropriately machined, they do not need to be bored again. These inserts are very wear-resistant and, furthermore, valve stem clearance to valve guide clearance can be minimal (0.05mm/0.0005in) yet they do not seize.

If the original guide bore axis is picked up correctly, the axis of the insert will be very close to the original guide axis but still expect to have to regrind the valve seats as it's impossible to be perfectly accurate.

Caution! When inserts are fitted to any guide the top and bottom of the guide have to be trimmed. On the top of each guide there is a small chamfer which is there to facilitate the easy fitting of the valve stem seal. If the engine reconditioner removes a bit too much material from the top of the guide during the trimming operation some, or all, of the chamfer may be removed and, as a consequence, it may prove difficult to fit the valve stem seals (they are damaged beyond use when fitted). Advise the engine reconditioner of this fact before work commences.

Refacing valves

The valves will usually require their seat surfaces to be cleaned up. There should be no actual distortion of the valve head or run out requiring more than 0.075mm/0.003in to totally clean the surface. Both inlet and exhaust valves require a minimum margin of 0.5mm/0.020in. If any valve has less than 0.5mm/0.020in margin on it, the minimum restoration that can be done is to grind the actual diameter of the valve until there is suitable margin or, if this (slight valve head size reduction) is not acceptable to you, replace the valve.

The inlet valve seat meets the back of the valve which is ground at 30 degrees and this forms an acceptable back edge shape but, after some regrinding, the ground valve seat surface may become larger than the recommended size. When this happens the back of the valve can have a further angle ground on it at, say, 25 degrees to reduce the size of the seat contact width.

If the valve head is not running true (bent or distorted) in the valve refacing machine, discard the valve and get a replacement. These valves will stand being cleaned up, but they will *not* stand having the concentricity of the valve head restored and the surface cleaned up as well. The only way to restore a valve which has been reground too much (sharp edge or little margin) is to grind the diameter of the valve head smaller. This is acceptable provided the seat width is reduced accordingly. Valve area is lost but this is not a concern on a moderately modified engine. The fitting of brand new valves is recommended.

Refacing valve seats

The valve seats are very hard and can only be successfully trued-up by grinding. To obtain accurate seats, the

Bare Sierra Cosworth '200' cylinder block.

SPEEDPRO SERIES

guides must be on size and the pilot (which locates in the guide) on size too. The seats are refaced with three angles (as per standard) which is, effectively, an inexpensive way of forming a radius type seat. The outer edge diameter of the 45 degree seat must measure the same as the valve's diameter.

The standard recommendation for the width of the seats is 1.5mm/0.060in for inlets and 2.0mm/0.080in for exhausts. For these naturally aspirated engines, the seat widths can be reduced, with reliability, to 1.0mm-1.2mm/0.040in-0.048in for inlets and 1.50mm/0.060in for exhausts. There are no problems with these sizes, provided good air cleaners are fitted to the engine's intake system.

It's recommended that the cylinder head be planed (0.5mm/0.002in) just to clean it up and guarantee that it is flat before the head is assembled. This will preclude the possibility of cylinder head sealing problems.

Once the cylinder head has been assembled, place it on a bench with the head gasket surface facing up and fill each combustion chamber with clean paraffin (kerosene). If the valve seating is correct no fluid will be able to pass between the seats and into the ports. If fluid is found to be leaking into a port that valve and/or seat must be reseated and the test done again. There is no point continuing with an

Natural aspiration Cosworth piston, piston pin and wire circlips.

The three piston rings that comprise a set for a single piston.

engine build until the valve seating is proved perfect.

SHORT BLOCK

For details of checking and testing cylinder blocks see the Pinto section of this book.

A 'stock' block with minimum overboring is best.

Do not use a sleeved block for any competition engine (and, preferably, any high-performance engine). If a genuine Cosworth block is found to have a crack in the bore, rather than sleeve it get another block. Cosworth blocks are the same as Pinto blocks for most intents and purposes. The substitution of a standard Pinto block for a genuine Cosworth block is not a backward step on the basis of strength for the application. Pinto blocks are inexpensive and readily available.

All cylinder block main bearing tunnel sizes and bearing inside diameters, clearances and crush *must* be checked – use the Pinto procedure but refer to Ford's Cosworth Sierra specifications. Note that main bearing bolts are torqued to 88-101Nm/65-75ft lb.

Piston clearance

For when the block is bored, the manufacturer of the particular piston set purchased will have provided information by way of a specification sheet, which will give a recommended piston to bore clearance. If the block is not going to be rebored, the clearance is checked and, if found to be insufficient, the bores will have to be individually power honed (as opposed to hand honed) until each provides optimum clearance for its piston.

As an example of measuring piston clearance, a forged Cosworth piston (which is suitable for natural aspiration) is 90.70mm in diameter and the standard bore diameter is 90.82mm. The recommended piston to bore clearance for competition purposes is 0.125mm/0.005in. Check every piston in the bore in which it is going to be fitted. The clearance can be checked by placing the piston into the bore (crown first and deep enough so that the bottom of the piston skirt is in the bore) and then inserting feeler gauges between the side of the piston skirt and the cylinder wall.

PISTONS

Forged pistons are readily available as direct replacements for the low compression Cosworth pistons, as found in all original equipment turbo charged engines. Cosworth Engineering make a 12 to 1 raised top piston which is a 'drop in fit' into these engines. Accralite also make 11.8 to 1 pistons for these engines which are also 'drop in fit' items. These two pistons have deep valve reliefs which suit the standard diameter valves offering plenty of piston to valve clearance, though not limitless clearance. Always check the piston to valve clearance once the engine has been assembled to make sure that there is enough clearance, especially when long duration camshafts are fitted.

Caution! Always fit brand new round wire piston pin retention circlips – *Never re-use old circlips*. Use plenty of oil on the aluminium piston pin bore when fitting circlips.

SIERRA COSWORTH & COSWORTH-HEADED PINTO ENGINES

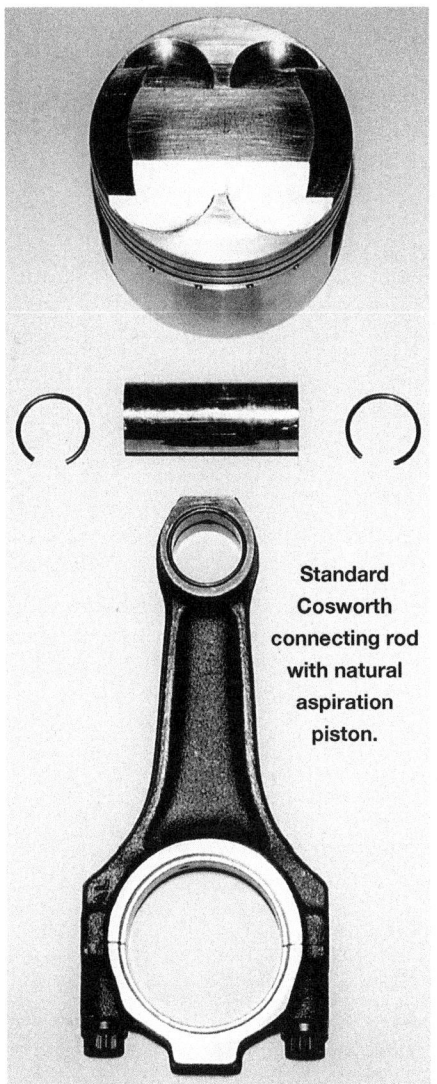

Standard Cosworth connecting rod with natural aspiration piston.

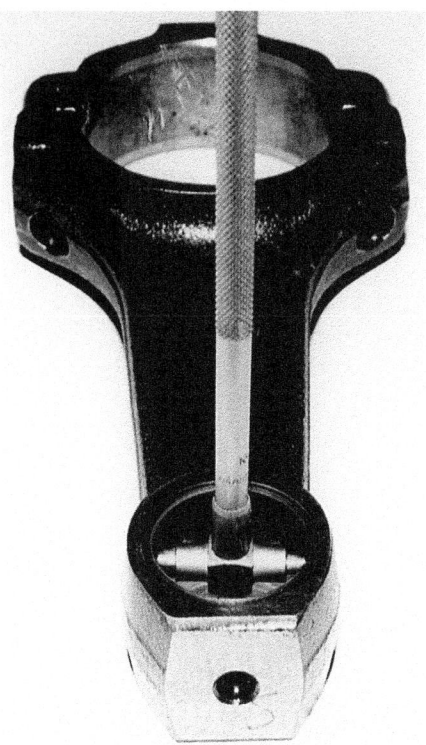

Small end/piston pin bush being measured with a telescopic gauge.

Inside micrometer being used to measure big end tunnel diameter.

Before the pistons are fitted to the connecting rods, read the piston pin oiling section to check whether the modifications suggested are applicable.

Piston weights must to be equalised (see Pinto procedure).

PISTON RINGS

Forged pistons almost always come with a 1.0mm thick top compression ring which is often chrome-plated. The second compression ring will be either 1.5mm or 1.75mm thick and will often be a 'moly' ring. The oil control ring will usually be a one-piece item with an expander behind it. The oil groove will have a series of holes drilled through into the underside of the piston crown, and there will often be four more holes drilled in the piston skirt below the oil ring groove to aid draining away of excess oil.

VALVE CLEARANCE

A Sierra Cosworth engine with, for example, Cosworth raised top forged pistons suitable for natural aspiration and a cylinder head that

Standard Sierra Cosworth connecting rod bolt.

How the connecting rod should be held in a vice (fitted with jaw protectors) to torque each of the nuts.

SPEEDPRO SERIES

has been planed 0.020in/0.5mm, will have approximately 8.0mm/0.320in between the seated inlet and the piston crown when the piston is at TDC. The exhaust valves, on the other hand, have 6.0mm/0.240in clearance between the seated valve and the piston's valve reliefs at TDC. This is very generous but, while the valves cannot collide with each other (by design), they can collide with their respective piston crowns through incorrect camshaft timing.

It's possible to bend valves when setting up the valve timing, but to do so the basic timing has to be out by a considerable amount. When the camshafts are changed for ones with longer duration and faster lift rates, the amount of leeway is reduced as the inlet valves are opened faster and earlier and the exhaust valves are closed slower and later.

CRANKSHAFT

Caution! Get the correct bearing shell inserts. Connecting rod bearing shell inserts *must* be heavy-duty Ford, Cosworth or Vandervell to preclude bearing failure. If in doubt, order them from a Ford dealer (ask for Sierra Cosworth bearings) or a specialist high performance parts supplier.

The crankshaft *must* be dead straight and *must* be crack-free – see the Pinto procedure.

The crank doesn't need to be detailed.

Crankshaft journals can be individually ground to provide optimum (within tolerances) bearing clearances.

CONNECTING RODS

Standard Cosworth connecting rods are very strong and virtually 'bullet-proof'. These connecting rods have bushed small ends for the floating wrist pins and 3/8in bolts. The I-section

Damaged piston pin tunnel.

Groove machined into the piston pin tunnel. This can be accomplished by hand or on a milling machine.

is immensely strong. They can be used just as they come. The standard connecting rods weigh approximately 720 grams. Note that Cosworth rods are slightly longer (0.060in) eye-to-eye than Pinto 2000 rods.

All connecting rods *must* be checked and measured as per the Pinto procedure.

If the piston pin bushes are worn replace them. The *maximum* permissible bush internal diameter is 24.025mm/0.9457in. Any more than this and the wrist pin will be too loose with excessive play quite apparent. The wrist pin diameter is nominally 24.00mm/0.9445in.

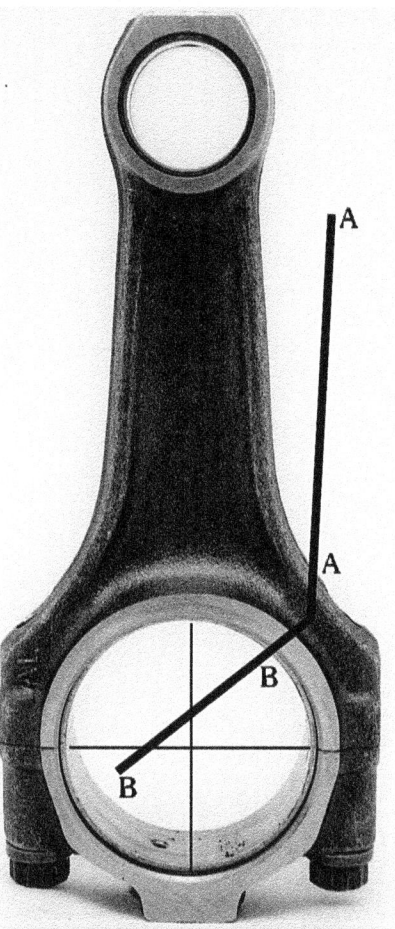

Line 'A-A' indicates the orientation of the 1.0mm oil spray drilling and 'B-B' the 2.0mm feed drilling. The intention is to spray the piston pin area within the piston so the angle of 'A-A' should, ideally, be between 3 and 7 degrees from vertical.

The big end tunnel diameter is nominally 55.00mm or 2.163in. On engines that have done a high mileage the big end diameter can be resized to ensure that the tunnel is perfectly round. Check new connecting rods just to be sure.

On used connecting rods, replace the bolts with brand new ones. The standard Sierra Cosworth bolts are adequate for any application.

Polishing the sides of the I-beams

SIERRA COSWORTH & COSWORTH-HEADED PINTO ENGINES

is not necessary, or recommended, but the forging flash partline can be polished smooth if preferred.

Connecting rod bearing clearance and crush

Use the Pinto procedures to check bearing clearances and crush. Sierra Cosworth connecting rod bolts are torqued to 55-60Nm/41-44ft lb.

With a set of, say, 0.25mm undersize bearing shell inserts fitted, expect the bearing internal diameter to be 51.77mm and, in relation to this, the crankshaft journal size should be 51.72mm. This gives a 0.05mm/0.002in bearing clearance which is ideal for these engines. Crankshaft journals can be individually machined (within factory tolerances), if necessary, to provide optimum clearances.

PISTON PIN OILING MODIFICATIONS

The information in this section will help prevent the frequent problem of piston pin to piston and connecting rod to piston pin 'galling' that is sometimes experienced with these engines. Competition, particularly racing, engines are susceptible to this problem. For racing engines the problem is preventable to a very large degree – even without carrying out this extra work detailed here – if the oil is heated and the pressure built up before the engine is fired. In many instances engines are started from cold and are immediately subjected to 2500-3500rpm which is no good at all for such an engine.

On dismantling many an engine will be found to have had piston pin galling problems, the piston pin bores in the piston will have 'picked up' and, often, the piston pin bush in the connecting rod will not look too good either. Essentially the problem is one of insufficient lubrication when the engine is started from cold with cold oil and the relatively high rpm that many engines are subjected to while warming up. The other cause of galling is lack of oil volume and/or pressure.

Part of the solution to this problem is to machine two grooves

Sierra Cosworth head gasket and bolt package.

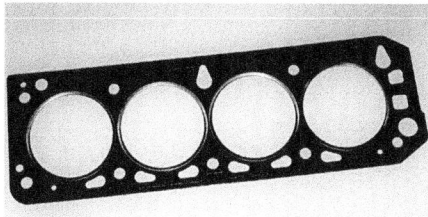

Group 1 head gasket.

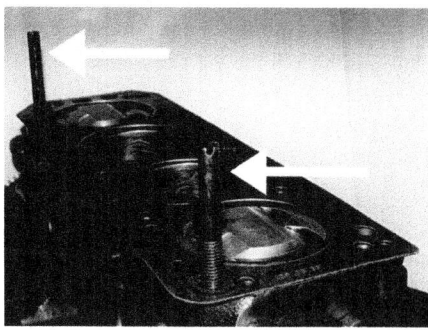

Two temporary studs to guide the head into correct position.

3mm/0.120in wide by 0.25mm/0.010in deep in the lower part of the piston's piston pin bore. The oil is then splash-fed directly around the underside of the piston and will travel into these grooves and lubricate the piston pin and piston bore. Having plenty of oil splashing around is all very well, but it has to be able to get to where it's needed and these grooves serve this purpose well. If new pistons are treated this way, and the rest of the procedure followed, galling damage is most unlikely. Some forged pistons have this feature as standard and it's common to find such grooves in many new standard pistons.

Front two pairs of camshaft lobes in the overlap position. Note how sprocket keyways are facing away from one another.

SPEEDPRO SERIES

The second part of the solution to the galling problem is to drill the connecting rods (see diagram) so that oil, from the crankshaft bearing, is squirted up the side of the connecting rod. The angle of the oil hole is *not* aimed at the lower edge of the piston skirt (common with many other engines) but, instead, straight up to the underside of the crown of the piston; the idea being that the oil is needed at the wrist pin bores of the piston and the piston pin end of the connecting rod. Extra lubrication of the piston skirt is not required or desirable (overloads the oil control ring).

For engines that get started with cold oil, having these extra jets of oil is ideal. The extra oil is not needed once the engine and its oil are up to temperature as there's then plenty of oil splashing around. Any road-going Cosworth engine should have these holes drilled in the connecting rods to ensure that there is always plenty of oil around the piston pin, little end bearing and the piston pin bores of the piston during warm-up.

Note that the standard Sierra Cosworth engine came fitted with a spray bar (fed directly off the oil pump) that sprays oil onto the underside of the pistons for piston cooling and piston pin lubrication purposes. This spray bar system is removed when the engine is dry sumped.

Caution! Because of the problems often experienced with piston pin lubrication, grooving of the piston pin bores is recommended for all engines (unless the pistons already have grooves).

Caution! The connecting rod oil hole *must* be added to all Cosworth Sierra engines which no longer have a spray bar (for whatever reason) and all Pinto blocks fitted with Sierra Cosworth rods (standard Pinto rods have suitable oil spray holes).

All of the Holbay crankshaft and connecting rod configurations for the Pinto engine apply to the these engines as well. Holbay supply pistons that suit the Cosworth cylinder head. Clearly it is quite possible to have a larger than standard Sierra Cosworth engine.

FITTING CYLINDER HEAD
Cylinder head gasket

Caution! The cylinder head *must* be completely flat and so must the block deck: remachine these two surfaces as necessary to ensure perfect flatness. If one surface is not dead flat, the chances are that the gasket will blow. For the block this machining is best done when the engine is completely dismantled: it's possible to mask a built-up short block and grind the deck but this method is *not* recommended. The deck of a block will usually have pulled up slightly around the stud holes, but it's usually only necessary to remove 0.05mm-0.10mm/0.002in-0.004in to totally clean up the block deck. It's a good idea to countersink all bolt holes too. Remove the *minimum* amount of material necessary from the cylinder head. Consider the maximum amount that can be removed to be 1.5mm/0.060in. No really useful purpose is achieved by planing the head, especially to gain compression: there are too many trade-offs to make this viable. The standard head is 139.0mm/5.472in in depth.

The best head gasket to use is the Group A Cosworth one. The Group A gasket is expensive, but will hold any compression pressure in a naturally aspirated engine. This gasket has cylinder apertures of 93.0mm/3.66in diameter, which makes it ideal for standard bores and overbores of up to 1.5mm/0.059in in diameter).

Caution! New head bolts *must* be fitted each time the cylinder head is replaced and torqued up as is the standard recommendation. The bolts are all tightened in three stages: 1) 20-25Nm/15-18ft lb; 2) 45-50Nm/33-37ft lb; 3) a final 180 degree turn.

The head gasket and head bolts come from Cosworth Engineering with the gasket sandwiched between plywood sheets for the absolute protection of the gasket during transit. *Do not* open the package until the head gasket is actually going to be fitted to the engine. This will avoid any possibility of damage prior to fitting.

Although the head fitting procedure is relatively straightforward, it's very easy to bend valves, especially if the camshafts have long duration and, more importantly, inlet opening points of 35 degrees or more and the exhaust closing points of 35 degrees or more. This potential problem ties into how deep the valve reliefs in the tops of the pistons are: it's less of a problem with forged deep valve relief racing pistons.

Caution! Before attempting to fit the head, make sure that *all* of the pistons are at least 8mm/0.31in down the bores.

The cylinder head is always installed with the front two pairs of camshaft lobes in the overlap position. This means that the valves of number one cylinder are fully seated as they would be at TDC on the firing stroke.

As for high-performance applications the standard non-adjustable drive sprockets are almost always replaced with adjustable ones, the standard timing marks are no longer available. The way around this problem is to pick up the same timing positions but by a different means: the keyways of the camshafts are ideal datum points for this procedure. With the two

SIERRA COSWORTH & COSWORTH-HEADED PINTO ENGINES

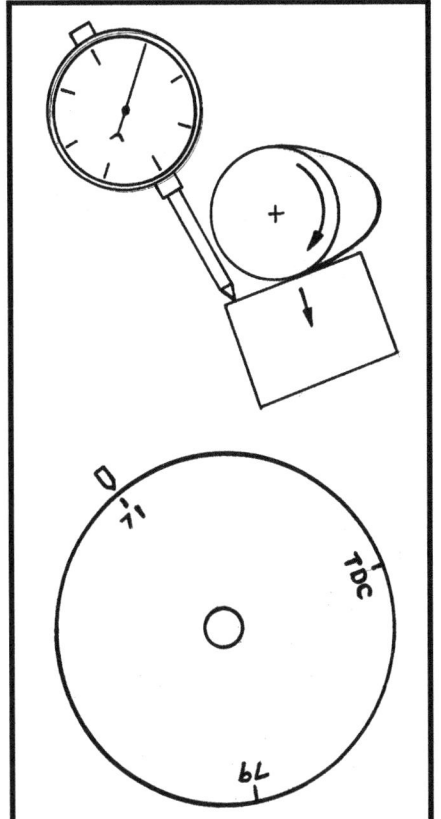
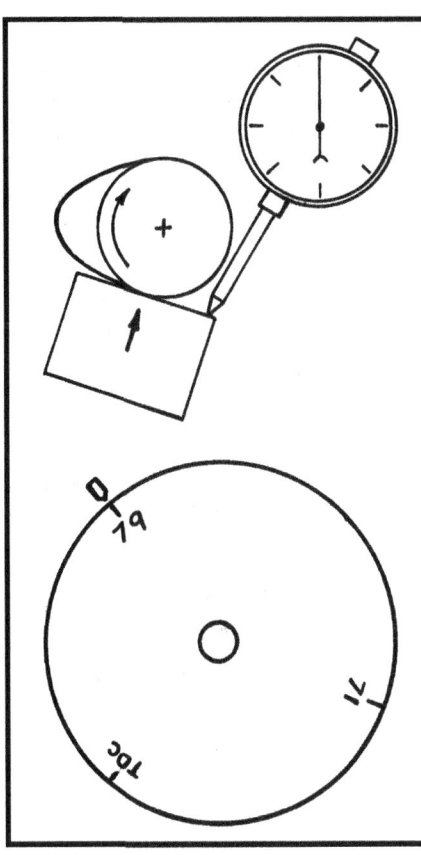

An example (L1 cams) of how camshaft timing is set. Left exhaust timing and right inlet timing.

Timing disc mounted on crankshaft.

camshafts installed in the cylinder head they are rotated until the keyways of each are facing outward and are in the horizontal plane: this procedure also confirms that the camshafts are each in their right side of the cylinder head.

It's far easier to correctly position the cylinder head on the block if the head is guided on. The easiest way to do this is to fit two long (at least 150mm) 12mm diameter studs, with screwdriver slots to facilitate their removal, one at each end of the block. The studs should have rounded ends so that the cylinder head is easily slid onto them: the head is simply lowered over the studs and down onto the block. The head is accurately positioned so there is no possibility of those valves that are open getting knocked as the head is moved around to line up the holes.

The two studs are removed after eight *new* stretch bolts have been loosely fitted to the cylinder head and then the last two stretch bolts fitted in their place.

With all bolts fitted but untensioned, move the cylinder head side to side and front to back to centralize the head as much as possible on the bolts and then tension the bolts to the correct torque and in the correct sequence.

At this point the cylinder head is correctly fitted, the retaining bolts fully tensioned and the valves of number one cylinder are in the overlap position. Using a 19mm open-ended spanner, turn the crankshaft, gently, so that the piston of number one cylinder is at TDC: it should turn easily and without locking (piston to valve contact). If adjustable cam sprockets are fitted, lock them in the middle of their range of adjustability. The timing belt can then be fitted and tensioned and the engine will be basically in time.

SPEEDPRO SERIES

After the belt has been fitted the engine can be turned via a 19mm spanner on the crankshaft using *only* hand pressure. Turn the crankshaft clockwise only (at this point) and *stop* turning *the instant* the crank locks. If it does lock, check the piston to valve clearance.

If the timing positions are effectively lost, remove the belt and see if the crankshaft can be turned anticlockwise. If it can't, leave the crankshaft where it is. Check to see which pistons are at, or near, TDC (obviously pistons nearer BDC are not involved), and, using the special tool, check to see which valves are in contact with the piston. Once this is known the direction in which to turn the camshaft will be clear (turning the camshaft one way will be pushing the valves harder, the other way will move the valves away from the piston). Start again with the number one cylinder piston 12mm/0.47in down the bore (approaching TDC when being turned clockwise) and both camshafts in the overlap position (see photo); turn the crankshaft until the piston comes up to TDC and then stop. Fit the drivebelt and tension it.

CAMSHAFT TIMING

The camshaft manufacturer will have supplied the full lift timing position and the opening and closing points of both the inlet and exhaust valves. Some figures, as supplied, will offer several timing options because tests have proved that particular positioning of the camshafts results in more power being developed. It pays to test the various recommended settings and, provided the engine has plenty of piston to valve clearance, there will be no accidents.

To time the camshafts accurately, a precise method of determining crankshaft degrees *must* be available (usually a timing disc or accurate marks on the crankshaft pulley).

With an engine equipped with adjustable 'vernier' camshaft sprockets, it's relatively straightforward to check and adjust the valve timing. The manufacturer or reprofiler of the camshafts will have supplied timing figures (full lift timing figures or opening and closing points) and, if the crankshaft pulley/timing disc is correctly marked, the camshaft timing should be set as recommended and the engine run and tested.

If the engine feels to be lacking in response, or not going as well as another example of the same engine that has similar specifications, check the camshaft timing. With the cam cover removed, use a dial test indicator to measure the precise point at which the cam follower stops moving down and zero the dial. This method of timing is equally well suited to full lift timing or absolute timing (as in the exact exhaust opening and inlet

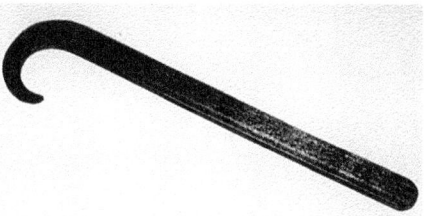

You can make a tool like this to lift the cam followers and valves.

closing points). Absolute timing points (the exact inlet closing and exhaust opening points) are almost always a more important consideration than the full lift timing point when timing an engine to deliver optimum power. Inlet closing at 80 degrees ABDC is the latest timing point to use and exhaust opening at 80 degrees BBDC is the earliest timing point to use.

In the final analysis, it does not matter what method is used to time the camshafts. What does matter is consistency of measurement and knowing precisely, in relation to crankshaft degrees, how camshaft events have been set. Once this

The follower lifting tool in use. The valves can be opened to a point where piston contact occurs.

SIERRA COSWORTH & COSWORTH-HEADED PINTO ENGINES

information is recorded, the engine can be run with several different camshaft timing settings (all within the allowable range) to gauge the performance obtained from each. **Caution!** Always check valve to piston clearances after each readjustment and *before* running the engine.

VALVE TO PISTON CLEARANCE – CHECKING

Checking the depth and position of *every* valve relief is *essential* to be sure there is sufficient clearance. Blocks and heads get planed for various reasons, different pistons get fitted, different camshafts (with high lift and long duration) get fitted, and so on, and all of these things effect valve to piston clearance. The only safe solution to this problem is to check the clearances with the engine assembled and, in so doing, take into account all of the variables that exist in any particular engine.

If insufficient clearance is found, it can mean a partial engine strip to increase the depth of the valve reliefs in the tops of pistons but this is preferable to bending the valves. Camshafts can often be set so that the pistons and valves will not collide, but may have to be moved so much that the optimum timing points are sacrificed. It's preferable to have as much piston to valve clearance as possible (as much as the piston crowns can tolerate without weakness). Any compression loss due to the removal of material to increase the depth of the valve reliefs can be disregarded as a minor consideration compared to the increased reliability of the engine.

A special tool which will allow the checking of piston to valve clearances irrespective of what has been done to the engine. This simple gadget hooks under the core of the camshaft and allows the camshaft follower to be moved away from the camshaft lobe by the rotational movement of the tool.

With the engine positioned at TDC for exhaust valves (closing) and TDC for inlet (opening) the tool can be used to lift the camshaft follower away from the camshaft lobe until the valve lightly touches the piston. Feeler gauges are then used to measure the gap that is formed between the camshaft lobe and the camshaft follower and this gap is the amount of piston to valve clearance available with the camshaft and piston at this particular point.

It's good policy to have as much piston to valve clearance as possible on both inlet and exhaust valves on these engines (you can't have too much, especially on the exhaust valves!). It is not really advisable to fit pistons that do not have four correctly positioned valve reliefs which are less than 4mm deep (inlet and exhaust). Long duration camshafts and, more specifically, camshafts with a lot of overlap are more critical with regard to valve/piston clearance than camshafts with less overlap.

Caution! When a new camshaft drivebelt is fitted its tension should be checked on a regular basis (every 500 miles/800 kilometres) because many new belts 'settle' and until the specified tension is maintained camshaft timing is variable. Once it is established that the belt tension has stabilised, checking intervals can be extended. Competition engines should have the tension of new belts checked after each race and, once it is established that the belt setting has settled, the interval can be increased to checking before each race meeting.

Every time the drivebelt needs to be retensioned, the camshaft timing should be checked. If the camshaft timing becomes retarded the exhaust valves will begin to lose their piston to valve clearance and contact could be the result (depending on how much piston to valve clearance there was originally). When the drivebelt loses its correct tension the camshaft timing will retard and, as the exhaust camshaft is furthest away from the front pulley (in belt terms) it will have

Rover/Honda HT wires fitted to Cosworth after minimal modifications to the sparkplug caps.

SPEEDPRO SERIES

Bosch Peugeot-type distributor cap with horizontal HT terminals. Note vacuum take-off (arrowed) in manifold for vacuum advance system.

retarded more than the inlet camshaft. Check the actual camshaft timing (as opposed to belt tension) on a regular basis, especially the exhaust valve closing point. Competition engines should have the camshaft timing checked before each event.

Caution! Failure to monitor the belt tension and/or failure to correct any subsequent alteration in camshaft timing, can result in piston to valve contact which will at least necessitate the removal of the cylinder head and the replacement of bent exhaust valves. It's *essential* that these engines have sufficient exhaust valve clearance. The inlet valves are not as critical in this sense because, if the camshaft timing becomes retarded, the piston to inlet valve clearance is increased.

IGNITION SYSTEM

The standard Bosch Pinto/Sierra distributor in either of its forms (electronic or contact breaker points) is ideal for use on this engine. The electronic type will use a higher output coil or one that is rated for electronic ignition systems. The points type ignition coil must be rated for points use or of ballast resisted type (the electronic distributor coil will burn the points out).

The distributor (points or electronic) *must* have 18 to 19 degrees of advance BTDC at idle (1100rpm) and a full advance of 32 degrees BTDC. The full advance is set to be 'all in' at approximately 3500-3700rpm.

The Bosch distributor has vacuum advance and, for road use, it should be retained as it will improve fuel economy. In the past it was common in high-performance applications to disconnect the vacuum advance and, frequently, to remove its mechanism from the body of the distributor. This was done because the vacuum advance mechanism is connected to the distributor's baseplate and, on some distributors, allowed too much fluctuation of timing advance. This problem does *not* apply to the Bosch distributor as it is of excellent quality. That said, competition only engines normally have the vacuum advance diaphragm canister and related parts removed and the baseplate brazed up to ensure no possibility of unwanted timing movement. When carburettors are fitted, the vacuum advance can be successfully operated by tapping into one of the engine's inlet tracts and running the usual small diameter pipe between it and the distributor.

High tension (HT) leads

Genuine Sierra Cosworth sparkplug leads are quite expensive but there is an inexpensive alternative set that can be used with only slight alteration. The leads in question are those aftermarket items designed for some Rover and Honda engines – the Rover 216 sohc, Accord 2.0, 2.0i (16 valve), Civic 1343cc and 1396cc (16 valve).

The alteration is confined to cutting the rubber away neatly around the cap to reduce the diameter so that the plug wire end fits down far enough into the cylinder head to connect correctly to the sparkplug. Use minimum resistance HT leads and for racing engines use copper core HT lead.

A further point is that for all general use Ford Fiesta sparkplugs are ideal (Bosch Super F6 DC, for example). For racing, NGK, for example, make special sparkplugs (eg:

Ignition and camshaft timing marks on crankshaft pulley illuminated and 'frozen' by a strobe light.

SIERRA COSWORTH & COSWORTH-HEADED PINTO ENGINES

Road going engine four-into-one system with 41.2mm/1.625in primary pipes of varying lengths (660-864mm/26-34in) which feed into a 50.8mm/2in main pipe.

NGK 9) but these colder sparkplugs will foul quite quickly in general road use.

The distributor cap to be used is the 90 degree HT lead terminal Bosch item that was used by Peugeots. The side exiting of the sparkplug leads lowers the effective height of the distributor and allows far more inlet manifold/carburettor clearance than the standard distributor cap.

Ignition timing
The critical advance settings for power on these engines in naturally aspirated form are 18 to 19 degrees at the idling speed of 1200-1500rpm and 32 degrees of total advance at approximately 3500-3700rpm. The amount of total advance remains the same for all engines and so does the rpm point at which maximum mechanical (centrifugal) advance is reached (32 degrees).

The vacuum advance system can be expected to increase spark advance by approximately 10 to 12 degrees when the engine is operating under vacuum at speed. When power is demanded from the engine, there is no manifold vacuum and therefore no vacuum advance being added to the mechanical advance.

Ignition timing checking & setting
With the crankshaft pulley correctly and accurately marked, it's a simple operation to check and then set, if necessary, ignition timing using a stroboscopic timing light ('strobe'). This is an *essential* piece of equipment if the timing is going to be set and maintained in the correct position.

The ignition timing can be checked to see exactly where it is at idle speed, then the engine revved to see exactly what number of degrees the advance goes up to and also, by checking the rpm, the engine speed at which full advance is reached.

The complete ignition system needs to be in excellent condition and the spark optimal at all engine speeds. You'll find more information in the Pinto section of this book. For those seeking in-depth information on ignition systems, distributor modifications, pulley marking and optimising ignition timing, a separate book in the SpeedPro series is available – *How To Build & Power Tune Distributor-type Ignition Systems* by Des Hammill.

EXHAUST SYSTEM
For best all-round performance use a four into one exhaust system.

For general high-performance road use the primary pipe diameter can be either 41.2mm/1.625in or 44.4mm/1.75in outside diameter exhaust tubing with the primary pipe length up to, but not more than, 965mm/38in long. The main pipe will generally be of 50.8mm/2in diameter.

For a competition engine which will operate continuously in the upper reaches of the rpm range (7000-9300rpm) 50.8mm/2in outside diameter exhaust piping approximately 711-762mm/28-30in long is used for the primary pipes. The main pipe will be either 54-57.15mm/2.125in or 2.25in outside diameter. The reason large diameter primary pipes are used is to move the torque peak up the rpm range. They do not increase the actual amount of torque that the engine produces, just move it higher up the rpm range where it is of more use for competition purposes.

The torque of an engine reaches a peak, usually in the mid range of the rpm band. This means that after the peak is reached it begins to diminish as the engine rpm increases. This situation is frequently shown graphically on engine power graphs. If, for example, an engine has 200 foot pounds of torque at 4000rpm, it

SPEEDPRO SERIES

may have 180 foot pounds of torque at 6500rpm and, perhaps, 165 foot pounds of torque at 7500rpm. If such an engine has primary pipe diameters of 41.2mm/1.625in and they are changed to 54mm/2.125in diameter primary pipes, the peak torque of 200 foot pounds may now be produced at 6000rpm instead of 4000rpm, 185 foot pounds at 8500rpm and, finally, 165 foot pounds of torque at 9000rpm. When engines do not have sufficient torque at high rpm they lose their 'urgency' – as the rpm gets higher, it feels as though the engine is losing power. Note, though, that what is gained at the top end is lost at the lower end of the power curve.

For all applications that will use a silencer, choose an aftermarket item suitable for the main pipe diameter and with minimum internal restriction.

CARBURETTORS

The usual carburettors are either 45mm or 48mm Dellortos or Webers (or similar) with 45s being ideal. The use of this type of carburettor is usually based on cost, known efficiency and reasonable simplicity of operation and maintenance. Very nice electronic fuel injection systems are available for these engines (good, but expensive) but still sidedraught carburettors seem to be the most popular option.

For those seeking in-depth information on jetting and setting up Weber DCOE and Dellorto DHLA carburettors for any application another SpeedPro book is available – *How To Build & Power Tune Weber & Dellorto DCOE & DHLA Carburetors, 2nd edition*, by Des Hammill.

Very nicely cast and machined aluminium inlet manifolds are available for these engines which bolt on to the cylinder head with only minor

Weber DCOEs on an inlet manifold designed for the Sierra Cosworth engine.

Underside of Webers, inlet manifold and cylinder head.

port matching being required. The inlet manifolds are usually cast to suit 45mm carburetors but, if 48mm ones are fitted, the casting can be opened out to suit. *Always* fit new MISAB spacers between the carburettor and the inlet manifold.

Road cars

Jetting will vary from engine to engine, but the following real life jetting example will suit a standard valve engine which has had the inlet and exhaust ports correctly ported. Further to this, the engine has 11:1 CR, 18-

SIERRA COSWORTH & COSWORTH-HEADED PINTO ENGINES

48 Dellortos on Sierra Cosworth manifold.

48 Dellortos on Sierra Cosworth head.

exactly 71 degrees BDC. The exhaust camshaft was also set with 0.015in instead of 0.010in valve clearance to reduce the total duration of the camshaft (reduces duration by about 5 degrees). These settings proved to be the best overall. The carburettors are twin 45mm Weber DCOEs -
38mm chokes.
155 main jets.
F16 emulsion tubes.
40 accelerator pump jets.
210 air correctors.
45F9 idle jets.
5.0 auxiliary venturis.
100 accelerator pump inlet/discharge.
200 needles and seats.

With this specification the production of very worthwhile power starts at 3800rpm. Below this rpm the engine is not 'fussy' at all and pulls off idle (1200rpm) through to 3800rpm at which point the engine gets very 'busy' as it works in the 'power band.' Once in the main power band (which starts at 6000rpm) there is nothing subtle at all about the delivery of power!

Competition cars

Quite a few of these engines have ended up being fitted into lightweight kitcars and sports cars of one type or another (and with good reason). One example has the standard Sierra Cosworth connecting rods and crankshaft while the pistons are 11.8 to 1 Accralites. The standard inlet and exhaust valves were retained but the cylinder head was re-worked as has been described in detail in this book. Maximum rpm was 9000.

The camshaft profiles used are the well-known BD4 (inlet and exhaust). These camshafts are timed on the basis of opening the exhaust valve (number one cylinder) at 74 degrees BBDC and closing the inlet

19 degrees of idle advance and 32 degrees of total advance 'all in' at 3500rpm. This Cosworth-headed 2000 engine has a Pinto IS block, so rpm is limited accordingly: the engine is occasionally taken to 7500rpm.

The camshafts (L1s) used have 306 degree duration and are phased at 47-79-71-55, with particular attention being paid to the inlet closing and exhaust opening degrees. This means that the inlet valve (number one cylinder) closes at exactly 79 degrees BTDC and the exhaust opens at valve at 78 degrees ABDC as opposed to being timed at the full lift position. The valve clearances were altered from 0.25mm/0.010in-0.30mm/0.012in for the inlets and 0.38mm/0.015in for the exhausts to get rid of a little bit of duration. The phasing is now effectively 55-78-74-55 instead of 56-80-80-56. This represents the maximum camshaft duration possible for best all round engine efficiency. Power is produced solidly all the way to 9000rpm with this set-up.

Idle advance is 20 BTDC at 1500-1600rpm and the total advance is 32 degrees BTDC and 'all in' at 3800rpm. There is no vacuum advance. The carburettors are twin 48mm DHLA Dellortos (45s go equally well). The car is fitted with -
40mm chokes.
165 main jets.
185 air correctors.
8011.1 auxiliary venturis.
7772.6 emulsion tubes.
40 accelerator pump jets.
7850.1 idle jet holder.
65 idle jet.
15mm shut off height float level setting.
25mm full droop float setting.

When driving this car it's a little bit difficult to comprehend the acceleration until used to it. The car is fitted with a close-ratio four-speed gearbox which feeds power via a 5.6:1 differential to the road surface through 10in wide slick tyres (21in diameter). The engine is rev-limited to 9000rpm and this gives a top speed of 104mph and speeds through the gears of 50mph in first, 68mph in second and 86mph in third. Power starts at 4000rpm with a second, better surge coming in at 6500rpm which then continues to the rev limit.

Minimal conversions
For engines which are Cosworth Sierra-headed (unported), but otherwise have minimal modification and have 'hydraulic' camshafts with around 280 degrees duration, 10.1-10.5CR and standard valves and springs, the following carburettor specifications will be a good starting point.

Weber 45 DCOEs.
36mm chokes.
F16 emulsion tube.
145 main jet.
40 pump jet.
45 F11 idle jet.
180 air corrector.
4.5 auxiliary venturi.

Dellorto 45 DHLAs.
36mm chokes.
8011.1 auxiliary venturis.
40 pump jet.
150 main jet.
180 air corrector.
7777.6 emulsion tube.
55 idle jet.
7850.1 idle jet holder.

Chapter 15
Starting engines & oiling requirements

Caution! It is recommended that all competition engines have the oil heated to 70C/158F *before* the engine is started. Most racing engines are dry sumped and it is relatively easy to use a paint stripping heat gun on the oil tank to heat the oil. The oil temperature gauge can be used to monitor the progress of heating. Alternatively, a permanent heater powered from an external 12v battery can be fitted to the tank. Oil temperature should not be allowed to fall below 50C/122F even after the engine has been started. Reheat the oil to 70C/158F before the engine is restarted.

Caution! Any new/rebuilt engine, or any engine that has been stood for a while, *must* be fully primed with oil before it is started: otherwise irreparable damage is done to the engine the instant it fires. The fact that the oil pressure soon builds up is no good at all.

To prime a dry sumped engine the belt drive of the dry sump pump can be removed and some form of auxiliary drive rigged up to turn the oil pump. For instance, a 12 volt automotive starter motor can be adapted to drive the dry sump pump via a toothed belt (not the same belt that normally drives the pump), or some form of direct drive to the front of the pump could be used. By this means hot oil is circulated through the engine and pressure built up as per normal while the crankshaft remains stationary. With the dry sump pump drive reconnected the engine can be started with confidence that the oil is warm, has been circulated and up to starting pressure.

A conventionally sumped engine which uses the standard type of oil pump can have the distributor removed and a modified distributor body (the spindle of which can be turned by an adapted automotive starter motor or an electric drill) substituted. The reason for using a starter motor is, of course, that a 12 volt battery can be the power source and priming of the oil system can be carried out virtually anywhere. The priming 'distributor' is modified to the extent that the drive gear is removed, the aluminium body turned off, all of the mechanical advance mechanism removed and some form of drive welded to the spindle. The spindle is easily lined up with the oil pump drive and there will no side loads imposed on the oil drive using this system. The ignition must be retimed afterwards, but this is a small price to pay. All this seems like a lot of trouble, but it's not half as much trouble as rebuilding the engine prematurely.

After priming, a competition engine is turned on the starter but not allowed to fire (separate switch for the ignition is required) until the engine has been turned over enough to circulate the warm oil sufficiently so that, when the engine is started, warm oil is present throughout the engine. Consider two ten second bursts on the starter with full starting oil pressure present as sufficient to make a proper start-up safe.

Also from Veloce Publishing ...

Ford Cleveland
335-Series V8 engine 1970 to 1982
Des Hammill

Years of meticulous research have resulted in this unique history, technical appraisal (including tuning and motorsports) and data book of the Ford V8 Cleveland 335 engines produced in the USA, Canada and Australia, including input from the engineers involved in the design, development and subsequent manufacture of this highly prized engine from its inception in 1968 until production ceased in 1982.

ISBN: 978-1-787110-89-2
• Paperback • 25x20.7cm
• 96 pages • 46 colour and b&w pictures

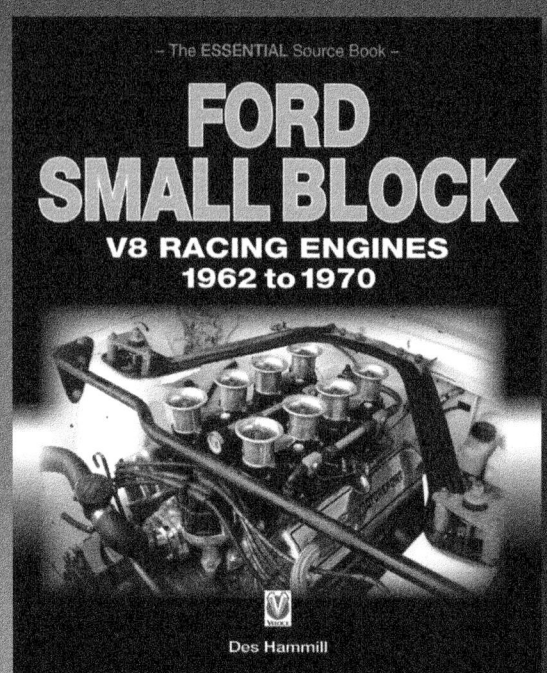

Ford Small Block V8 Racing Engines 1962-1970
Des Hammill

This book gives a rare insight into the confident, logical approach of Ford's Engine & Foundry Division engineers during the 1960s racing programmes. Few people know the facts about Ford's design strategy during this period – a time when outstanding technical decisions led to many events being won with larger derivatives of the 221ci Small Block V8 production engine, introduced for Ford's 1962-model year Fairlanes and Mercury's.

ISBN: 978-1-845844-25-7
• Paperback • 25x20.7cm
• 112 pages • 125 b&w pictures

For more details visit www.veloce.co.uk or email info@veloce.co.uk

Index

Advertised valve lift 79
Aftermarket aluminium flywheels 21
Aftermarket connecting
 rods 18-21, 40
Aftermarket replacement
 pistons 15-20, 23
Air filters 110
Altered rocker geometry 82-86
Alternative Ford connecting
 rods 16-19
Alternator 100, 101
Aluminium flywheels 21, 95
Auxiliary shaft bearing (replacement
 parts) 25
Auxiliary shaft gear wear 13, 14
Auxiliary shaft thrust plate 29

Back chamfering valves 56, 57
Balance (engine) 95
Balancing (equalising pistons/
 gudgeon pins) 95
Ballast resistor coils 99, 100
Block and main bearings 31, 32
Block and main bearings (inspection
 and checking) 31-35
Block planing 63

Camshafts (Cosworth heads)
 Camshaft followers 123-125
 Camshafts (high lift) 118, 119
 Camshafts (long duration) 118, 119
 Camshafts (low lift) 117, 118
 Camshafts (valve lift
  limitations) 116, 117
Camshafts (Pinto heads)
 Camshaft bearings 13, 24
 Camshaft kits 24
 Camshaft lobes and rockers 12, 13
 Camshaft pillars 13
 Camshaft spray bars 24
 Camshaft thrust plate 29
 Camshaft timing 71, 72
 Camshafts 66-72
 Camshafts (racing) 70, 71
Carburettors 107-110
Carburettor summary 109
Choosing camshafts 67-70

Clutches 21, 22
Combustion chambers
 (modified) 52, 58, 59
Combustion chambers
 (standard) 51, 52
Competition valves 26
Compression ratios 61-65
Condenser 98
Connecting rod bearing crush 41, 42
Connecting rod 'big end' bearing bore
 sizes 41, 42
Connecting rod 'big end' (optimum
 sizes) 42
Connecting rod 'big end'
 resizing 38, 39
Connecting rod 'big end' tunnel size
 (checking) 37, 38
Connecting rod bolts (alternative) 40
Connecting rod bolts (checking for
 'stretch') 40, 41
Connecting rod bolts (new) 39
Connecting rod bolts (removing old
 bolts) 38
Connecting rod centre to centre
 distance check 36
Connecting rod checks 35-42
Connecting rod 'check fitting' (onto
 crankshaft) 45, 46
Connecting rod crack testing 35
Connecting rod hardness (little
 end) 36
Connecting rods 'little end' to gudgeon
 pin fit 36, 37
Connecting rod straightness testing 35
Contact breaker points
 (distributor) 97, 98
Cosworth engine camshaft
 timing 134, 135
Cosworth engine carburettion (side
 draught Webers and
 Dellortos) 138-140
Cosworth engine exhaust
 systems 137, 138
Cosworth engine ignition 136, 137
Cosworth engine ignition timing 137
Cosworth engine (piston to valve
 clearance checking) 135, 136

Cosworth connecting rod preparation
 129-132
Cosworth cylinder heads
 Cosworth cylinder head fitting
  procedure 132-134
 Cosworth head to Pinto block
  camshaft drive modifications 114
Cosworth exhaust porting 116
Cosworth forged pistons 128, 129
Cosworth inlet porting 115, 116
Cosworth piston pin oiling 131
Cosworth piston ring sets 129
Cosworth piston to bore
 clearances 128
Cosworth porting criteria 114-116
Cosworth rebuilding 125-128
Cosworth valves 118-120
Cosworth valve guides 125-127
Cosworth valve refacing 127
Cosworth valve seat refacing 127, 128
Cosworth (valve spring pressure
 testing) 123, 124
Cosworth (valve springs for these
 heads) 120-123
Cosworth (valve stem seals for these
 heads) 120
Cylinder block 30
Cylinder block freeze plug
 retention 46, 47
Cylinder head (exhaust port
 sizes) 55, 56, 58-60
Cylinder head (inlet port sizes) 54, 55
Cylinder head (inlet valve guide
 alterations) 55
Cylinder heads (IS Sierra) 52, 53
Cylinder head planing 61-63
Cylinder head (port
 terminology) 48, 49
Cylinder head (valve
 unmasking) 54-56
Crankshaft bearings 24
Crankshaft checks to make 43
Crankshaft crack testing 43
Crankshaft detailing 44
Crankshaft front pulley degree
 marking 102-104
Crankshaft journal sizes 44, 45

SPEEDPRO SERIES

Crankshaft regrinding 45
Crankshafts (standard) 20, 42, 43
Crankshaft straightness testing 43, 44

Dellorto carburettors (jetting) 107-109
Distributor 96-99
Distributor cap 98
Distributor end float 97
Distributor drive gear 97
Distributor spindle 97
Dowelling flywheels 20, 21
Dry sumps 28

Electronic coils 100
Engine balance 95
Engine starting (from rebuilt) 141
Exhaust manifolds (primary pipe diameters and lengths) 91, 92
Exhaust manifolds (types) 90, 91
Exhaust ports (modification of) 57, 58
Exhaust system (Pinto) 89-92
Exhaust valve guide boss modification 56

Flywheels 20, 21
Flywheels (altering standard ones) 93, 94
Flywheels (steel ones) 94
Fuel pressure 110
Fuel supply 110

Gaskets (cylinder head) 63

Heavy duty timing belts 29
High-performance camshafts 66-70
High tension requirements for good ignition 101, 102
High tension wires 100

Ignition 96-105
Ignition coils 99, 100
Ignition summary 105, 106
Ignition switch 101
Ignition timing markings 102-104
Inlet manifolds 109, 110
Inlet ports 57

Large capacity engines 21
Large valve modification criteria 58, 59
Lash caps (correcting rocker geometry) 83
Loose sprockets 14
Low voltage coils 99

Main bearing bores (checking sizes) 34, 35
Main bearing cap register fit and fitting 31, 32
Main bearing caps 32
Main bearing 'crush' 35
Main bearing shell 'check fitting' 33-35
Main bearing shell insert location tabs 32
Main bearing tunnels (checking sizes) 32, 33
Module (electronic for distributor) 98

Oil pump drives 27
Oil pumps (standard type and up-rated) 27
'O' ringing blocks 64

Permanent ignition advance degree markings 102-104
Permissible bore oversize 30, 31
Pinto & Cosworth blocks (compression ratio) 113, 114
Piston pin oiling modifications (Sierra Cosworth) 131, 132
Piston pin to connecting rod 'little end' fit 36, 37
Piston rings 16, 20, 24, 129
Pistons 15-21, 23, 24, 30, 31, 63, 64
Piston to valve contact 13
Porting tools 54
Primary exhaust pipe diameters 91
Primary exhaust pipes 91, 92
Problem areas (Pinto) 10-14

Racing valves 25-27
Raised top pistons 63, 64
Ram pipes 110
Refacing connecting rod and cap matching faces 39
Removing connecting rod bolts 38
Repaired blocks 31
Rev-limiters 105
Rocker and rocker geometry (checking) 86-88
Rocker geometry checking tool 85-88
Rocker geometry (correcting) 82, 83
Rocker geometry (standard criteria) 81
Rocker geometry (when no longer standard) 82, 83, 86-88
Rocker sizes/designs 83-86
Roller camshafts (Holbay) 69, 70
Roller rockers 71, 86
Rotor arms 98, 99

Seals (block/cylinder head) 25
Seals (valve stem) 27
Side draught carburettors (Weber and Dellorto) 107-110
Spark plugs 100
Spark quality (checking) 100-102
Special bolts 25
Standard camshafts 66
Standard connecting rods 10, 11,
Standard pistons 15, 16
Static ignition advance 104-105
Strobe light ignition timing 105
Sump 28

Timing belt 28, 29
Top dead centre (TDC), the checking of 103
Total ignition advance 105

Useful addresses 9

Vacuum advance 105
Valve guides 14, 60
Valve lift (See Advertised Valve Lift)
Valve retainers and collets/keepers 27, 75, 77
Valves 25-27, 49-54
Valves (benefit of larger ones) 50
Valves (competition) 50-52
Valves (Group 1 type 1800-2000cc engines) 50, 51
Valves (Group 2 type inlet valve 2000cc engine) 51, 52
Valve size summary 53, 54
Valve spring data (standard) 75-77
Valve spring dimensions 74, 75
Valve spring fitted heights 79
Valve spring poundages (measuring) 77-79
Valve spring retainers (Titanium) 77
Valve springs (Pinto) 73-80
Valve springs (Sierra Cosworth)
Valve spring (summary) 80
Valve stem length (correcting rocker geometry) 82, 88
Valve stem seals (replacement parts) 27
Valve throat and port modifications (standard sized valve) 56
Valve throat sizes 54
Valve unmasking 57

Weber carburettors (jetting) 107-109
Wiring and connections 100

Printed and bound by CPI Group (UK) Ltd, Croydon, CR0 4YY
22/03/2026
02076211-0001